Succeed in Science and Avoid Getting a Real Job

by Lloyd D. Fricker

featuring cartoons by Nick D. Kim

ISBN 978-1-4664-2517-0

Printed by CreateSpace.

https://www.createspace.com/3510904

Table of Contents

Succeed in Science

Introduction

So you want to be a scientist running your own lab. Or maybe you already are, and want to know where you took a wrong turn to end up in this crazy profession. If so, read on—hopefully you'll find something that will make you smile. You may also find some advice that will help you move up to the next level of the career track. After all, many of the essential things you need to succeed in science are not typically taught in school.

This book started out as a collection of humor pieces and evolved into a book that includes serious advice for junior scientists hoping to succeed as independent scientists. It is intended as an overview of the process of finding a job as an assistant professor in academia (or a related position in a research institute), setting up a lab, and moving up the ranks. Some chapters include many humor pieces and cartoons while others are mostly serious advice describing essential things for all scientists, such as writing scientific papers and attending meetings. Some chapters are specialized for academic positions that involve teaching and writing grant applications. It is not possible to completely cover any of these areas in a single chapter; entire books have been written on each of these subjects. Instead, this book is intended as an overview, with humor, and with some references for further reading.

You can read this book cover to cover or browse it for topics that are relevant to your particular interests. Or just thumb through for the cartoons and humor pieces. The distinction between humor and serious advice in this book is not always clear; most of the humorous pieces (including the cartoons by Nick Kim) have a serious point to them, and much of the serious advice is presented in an irreverent style.

Succeed in Science

Chapter 1
What It Takes to Succeed

Chances are you love science—otherwise you wouldn't be reading this book. When you were younger, you probably enjoyed studying science. And at some point you did an experiment and got hooked on the thrill of discovery. You found out that science wasn't just something that you studied, it was an activity—one that involves exploration and creativity. You made the transition from reading about ancient discoveries in textbooks to actually *doing* science. And this was fun.

There is a similarity between science, art, music, and sports—all of these activities are more fun to do than to study. Kids don't start off *studying* art, music, or sports—having to memorize all the history before they can pick up a crayon, sing a song, or kick a ball. Kids just *do* these activities for fun! Then, if they are good at it and want to become professional artists, musicians, or athletes, they have to spend time studying the subject. But the way science is currently taught, the studying comes first. You have to memorize lots of details and do well on your exams, and then you finally get a chance to do real experiments. Research is a whole different experience from studying, and you were lucky enough to be exposed to it. And you love it! Now you want to move to the next level and run a laboratory of your own.

To do experiments well, you usually have to be good with your hands, be able to focus, and be persistent in order to follow through to the end of a project (or at least a publication, if not the end). When you get to be a lab director you have to learn to step back and focus on the big picture, keeping ten things going at once. You also have to be able to recognize when to drop a project that's not progressing. In addition, you need to learn how to hire the right people and motivate them, get funding for your lab by writing successful grant applications, publish papers in good journals, balance your budget, participate on committees, and teach. As you advance, you also have to contribute to your field by reviewing manuscripts and grant applications, and by helping with the organization of scientific meetings. In most cases, there is little training for any of these tasks, so the real learning occurs once you actually have to do these things. Many of these topics are covered in later chapters in this book.

It is helpful to know what you are getting into before you make the leap. It is also a good idea to consider other options—just because you trained as a scientist doesn't mean that you have to make the transition to assistant professor in academia or an equivalent lab director position in a research institute. There are many potential jobs for people with training in science, many with better pay and shorter hours than a job running a research group.

To survive as the director of a laboratory takes an extreme dedication to science. Is science your job—something that pays the bills and keeps you off the street, or is it your passion? Do you wake up in the morning thinking about your experiments, eager to get to lab and test your ideas? There are also other things needed for success in science—enthusiasm alone will not be enough. The following section describes the factors affecting success. Some of these reflect innate traits while others are things which can be taught (and are described in later chapters).

The first three of the following points are essential traits for successful people in a number of professions, not just science. The other points listed below are important in science (as well as other fields), but are not essential—there are examples of highly successful scientists with deficiencies in one or more of these areas. However, if you are weak in many of these areas, it may be better for you to seek a career that does not involve running your own laboratory.

#1 – Motivation

Without a passion for science, it would be very hard to succeed in academia as a lab director. Long hours are needed, and there will be times when experiments aren't working as planned, your submitted manuscripts and grant applications are rejected, and you get scooped by a competitor (sometimes all at the same time!). Unless you have a strong inner drive to be persistent and keep going, it will be hard to succeed. While long hours are needed, this doesn't mean you have to devote every waking hour to science—only some scientists do this. It is possible to have a life outside of the lab and be successful in science. Of course, the social life of a scientist is smaller than that which is possible for other professions, but at least some social life is possible. And if you count time spent at scientific meetings as your social life, then you can have an amazing social life as a successful scientist—one that involves trips to distant places and a large network of international friends.

#2 – Intelligence

Without intelligence, it would be hard to succeed in science. However, you probably are smart enough. Sure—we all want to be smarter, but the most successful scientists are generally not any smarter in the classic sense than less successful scientists. It is not necessary to score at the super-genius level on standard IQ tests in order to make it as a scientist—being above average is sufficient, and if you have a Ph.D., you are above average in intelligence.

#3 – Creativity

Creativity is distinct from intelligence and much more variable among scientists. Both creativity and intelligence are essential for success. You need to be able to think outside the box and come up with new ideas to explain results that don't appear to make sense. When you are a junior scientist in training (i.e. student, post-doc), you can often get by with creative advice from your mentor and other advisors, but to succeed on your own you need to be creative.

#4 – Common Sense

In academia, common sense is not all that common. The classic stereotype of an absent-minded professor is based to some extent on reality, although it is rare to see a real scientist as extreme as the stereotype typically portrayed in movies. Common sense is not the same as intelligence or creativity, and a combination of all three is especially powerful. Creativity gives you ideas, intelligence allows you to develop the ideas, but common sense is needed to determine which are most likely to succeed. According to Linus Pauling, a two-time Nobel Laureate who certainly had many good ideas, "the way to get good ideas is to get lots of ideas, and throw the bad ones away." Common sense allows us to recognize the bad ideas.

#5 – Communication Skills

The ability to write clearly and distill complex ideas into comprehensible text is essential for all scientists who need to publish papers and obtain funding for their research. It is also important, although not as essential, to have good oral communication skills and be able to give an excellent lecture to a roomful of scientists or students.

#6 – Networking Skills

Most fields of science involve teamwork, and for this it is necessary to set up collaborations and work together. It is also important to have a group of colleagues you can ask for advice on important matters such as grant applications. Developing a network of colleagues involves social skills.

#7 – Memorization Skills

To get through college and graduate school, you needed to memorize many details. You will also need these skills later in your career. While the ease of looking things up on the internet has made memorization of facts less important than it once was, you need to remember enough to be able to find the relevant facts. And, in the process of coming up with good ideas, it is useful to remember bad ideas of the past. As said by the philosopher George Santayana, "those who cannot remember the past are condemned to repeat it." Repeating bad ideas is a complete waste of time.

#8 – Open-minded attitude

Although not as important as motivation, another personality trait that impacts on success is that of attitude—whether you have an open mind and can accept things that don't fit your hypothesis. While it may seem that this would be essential, there are examples of scientists who insist that data support their ideas and ignore anything that says otherwise, even when the evidence against their theory is strong. While persistence is important, it is dangerous to ignore the possibility that your favorite theory is wrong. Long-term success requires that your theories hold up over the years, so keep an open mind.

#9 – Opportunity, resources, and luck

In addition to the above traits that will help you succeed in science, there are three other essential factors which are generally beyond your control: opportunity, resources, and luck. Even the most brilliant, creative, and motivated person cannot succeed if he or she is not given the chance with adequate resources. Funding is difficult to get, and once obtained, never seems to be enough. Luck is definitely needed to get funding; the grant application process has considerable randomness and is never guaranteed, even for highly successful scientists. There is also an element of chance to the scientific discovery process—will you find something important and be the first to publish your finding, without being scooped by a competitor? Although luck is not something most people think they can control, there are some steps one can take to increase the odds of success. These are described in a later chapter in this book. But for now, to round out this chapter, it's time for a humor piece!

Famous Scientists as Role Models

It has been said that great men stand on the shoulders of giants. So for all of you standing on the shoulders of midgets, get down this instant and find a giant to stand on. Of course, the hard part is to figure out what to pay attention to—we're all bombarded by advice, comments, rantings, etc, and it takes a true genius to figure out what is important and what isn't.

In considering a career in science it is worthwhile to take a look at one of the most famous scientists of all time: Albert Einstein. After graduating from school, Einstein first tried his hand at tutoring but his poor students were bewildered at his lengthy explanations, and his tutoring business was slow. He got a job as a patent clerk but also didn't last in this endeavor.

Where did Einstein get his ideas? Maybe it wasn't the shoulders of giants after all. Perhaps it was just everyday people who gave him the ideas in the first place. It is well known that Albert Einstein developed some of his most important theories while working as a patent clerk in Zurich, Switzerland. One of his most important theories of all time is simply stated as "ein job mit der governmenten allowen grossen schnoozen" which is roughly translated as "government jobs leave one with lots of time on one's hands." Einstein left this job when given an academic position because of his second fundamental theory which is loosely translated as "a job as a professor sure beats working for a living." In addition to these theories, Einstein came up with a bunch of other stuff while working as a patent clerk. Some historians have argued that Einstein's best work was done while employed in the patent office. However, nobody has previously attempted to analyze the potential inspiration offered by this government job.

It is clear that an analysis of the writings of the young Einstein as a patent clerk would be revealing of his developing mind. What patent applications did the young Einstein read and evaluate, and how did this influence his later career? To save time and money, I have used telepathic dreams to learn the content of Einstein's writings as a patent clerk. The following are the visualization of those writings.

Invention: Ein Mousentrappen (a mouse trap)

Summary: This purpose of this device is to catch mice. The mice are lured to the device by a cube of cheese. Once they grab the cheese, a steel spring is released and the mouse is instantly killed.

Evaluation: This is a potentially important device but too little attention in the application is devoted to the theory. It is unclear whether the device will have a suitable efficiency to be useful in controlling the rodent population. The efficiency of the device (E) is a function of the number of mice in the house (M) and the size of the cheese cube (C^3) by the following equation $E = MC^3$.

Recommendation: A device that targets mice solely based on their genetic composition (i.e. the fact that they are mice) is socially unacceptable. Instead of killing the mice, a device that allows for their capture and rehabilitation would be more socially appropriate.

Invention: Ein Wristenclocken (a wrist watch)

Summary: This device combines the well known pocket watch with the recent development of an electrical clock. A small electric generating device known as a "battery" will be used to produce the electrical current to run the device.

Evaluation: This application does not adequately cover the theory behind an electrical wrist clock. In order to function properly, the battery must have sufficient energy (E) to move the clock hands around the watch dial. Taking into account the mass of the hands (M) and the circumference of the watch dial (C), the required energy is described by the equation $E = MC$.

Recommendation: Battery power will never be sufficient to run this device. Instead, the clock should use wires to connect to a power source. An advantage of wires is that this would provide a useful method for travelers to find their way home (i.e. by retracing the wire).

Invention: Ein Pocketknifen (a pocketknife)

Summary: This is a simple modification of a knife in which the blades are foldable into a central handle. It is proposed that this knife will be used by the Swiss Army.

Evaluation: It is not clear why a modern soldier would require a knife that is not effective in combat, and appears better suited for cutting meats and cheeses at a picnic. The effectiveness (E) of this knife can be described as a function of the cross section of meat (M^2) and cheese (C^2) that must be cut, using the equation $E = M^2C^2$.

Recommendation: While the basic premise of a foldable picnic knife is an important advance and worthy of a patent, it is inappropriate to refer to this as a Swiss Army Knife since its role as a weapon in a modern army is dubious.

Invention: Ein Flashenlighten (a flashlight)

Summary: This device is a combination of the electric lamp with an electrical battery that allows for mobility of the light.

Evaluation: A wireless lamp with the light of many candles, but which is much safer than a torch, is an important advance. However, the light energy given off by such a device (E) will be limited by the mass (M) and charge (C) of the batteries by the equation $E = M^2C$.

Recommendation: This device will revolutionize night time travel. It may also be useful for theoretical physics experiments. For example, if one imagines traveling at the speed of the light from a flashlight (C), there would be a slowing of time and an increase in mass of the observer, such that the mass (M) would be related to the energy (E) by some kind of equation.....

Chapter 2
Finding a Job—the Application

It has been said that finding a good job is a real crap shoot, although why anyone would shoot crap is beyond me. After all, it isn't like the crap is attacking and you have to shoot it in self defense. All of the crap I have seen is pretty inert, so it makes little sense to go out and shoot it. Then again, perhaps this analogy is appropriate because much of the hiring process makes little sense. First, you're judged by your CV, which usually reflects a combination of the lab you're in, random luck, and to a small extent, your own personal abilities. But from the perspective of a committee reviewing hundreds of job applicants, this is the only practical way to sort out those who are most likely to succeed—the committees don't have the time or money to interview every applicant. A candidate who has been productive as a student and post-doc is likely to be successful later in his/her career, so it makes sense for the search committee to select those with the strongest CVs.

While a strong CV and cover letter are important, these do not guarantee you an interview. Recruiting committees are typically looking for someone in a particular area and this may not be explicitly stated in the advertisement describing the position. You may have an outstanding CV but not get an interview because your area of research doesn't fit the specific research area wanted by the committee. Most places want to have a balanced research group working on a range of projects that have some synergy between them but don't overlap too much. Who's in the department is an important consideration (more on that later). But first we must consider the basics: the application, the interview, and the human sacrifice. OK—so I'm just kidding. There's no human sacrifice involved. At least in most cases.

Finding a Job Opening

Start by deciding what type of job you are looking for, such as an assistant professor in academia, a scientist in a company or government laboratory, or something else entirely. If you are considering the academic route, think about what fields you are qualified for based on your post-doctoral research and previous studies. Then, take a piece of paper and write down your three top locations to live. Next, write down the salary you would like to make. Finally, take this piece of paper, wad it up into a ball, and toss it into the garbage. You'll never find a position like that! Instead, you have to settle for what is available and reasonably close to your area of training.

In the old days we would read job ads in the back of Science or Nature. Nearly all of these would be for the wrong level of position, either too senior (director of a department) or too junior (yet another post-doctoral position). The few jobs that were for positions at the assistant professor level were almost always a completely different field, country, or both. When we would spot a position that was vaguely related to what we studied and within a few thousand miles of our ideal location, we would get very excited even though it was quite a stretch scientifically. Now, with the advent of computer searches there is not the same sense of excitement as in the past—no buildup of multiple hours of relentless searching, week after week with no end in sight. It's just a few mouse clicks and keystrokes before you realize that despite many years of advanced training, you are not prepared for the vast majority of jobs. The following song sums up the job search.

The Job Ads Found in <u>Science</u>
(to the tune of "The Sound of Silence" by Paul Simon)

Hello job ads my old friend,
I've come to search you yet again.
Until I find a job that suits me well,
And lets me escape from this post-doc hell.
So I apply each week with my CV, hopefully,
To every job ad found, in Science.

The field of research matters not,
Or the degrees that I've got.
I apply to each and every place,
With an opening 'cuz just in case,
They could be looking for someone like me, possibly,
I cover all the ground, in Science.

Fools, thought I, why can't you see,
My innate creativity?
Although my résumé is rather slim,
A Nobel Prize I'll almost surely win.
I even put this down on my CV, the prize to be.
But still I'm searching round, in Science.

Ten thousand times I have applied
But not a one has yet replied.
Except to say quite simply "we'll call you."
But in the end I know they never do.
Why can't I get a single interview? Or a few!
But all I hear is the sound, of silence.

Give me a chance and you will see,
A dedicated employee.
I'll teach whatever courses that you need,
And gladly sit on any committee.
Until I find that tenure's been denied. First I'll cry.
Then I'll search around in Science.
{ending} I'll search the job ads found in Science.

The Curriculum Vitae

What is a CV but a résumé on steroids? Résumés are advised to be a single page and for "normal" jobs (i.e. not science-related jobs) applicants are advised to limit their CV to two pages. But a two page CV for an academic position would promptly be tossed into the recycling bin. A CV for a normal job will list objectives (i.e. to get a job doing essentially nothing all day and which pays large sums of money), and can even include hobbies. In contrast, a CV for an academic position usually doesn't bother with obvious things like objectives (to do science!) and hobbies (to do more science!). Instead, you need to fill several pages with your educational background, honors and awards, and most importantly, publications. The following are standard items that go in a CV.

<u>Definitely include</u>

Where you went to school (starting with college)

Employment history

Gaps in education/employment if there is a great reason (took time off to raise children; was abducted by aliens and held captive in their spaceship)

Publications and future publications (i.e. manuscripts definitely accepted for publication)

Likely future publications (i.e. manuscripts that have been written and submitted for publication)

Invited publications (book chapters, if any)

Invited seminars (if any)

Teaching portfolio (classes/courses taught, if any)

Consider including if you need to fill out your CV

Oral and poster presentations at meetings (include references to meeting abstracts if they were published)

Possible future publications (i.e. manuscripts you are currently writing). Not all search committee members want to see these, and some people will be turned off by a long list of manuscripts that are "in preparation." However, if you really are writing a manuscript (and not just planning on writing once the experiments are completed) then it is OK to include one, or at most two of these "in preparation" manuscripts.

Do not include under any circumstances

Where you went to preschool, kindergarten, and grade school

Any honors or awards in the above school years unless really impressive (Intel finalist, Presidential Medal, Nobel Prize, etc are good to include, but omit the gold stars you got for your third grade science fair project)

The fact that due to your split personality, if you are hired, your employer will get two employees for the price of one!

Description of Research Interests

Most academic employers request a brief description of your research interests and/or future plans, whereas this is extremely rare for jobs in industry. There are ultimately two objectives of your research description: one is to see if you will be a good fit for the department, and the other is to judge the likelihood that you will be funded. Most academic jobs require grant support to fund the research program and a portion of your salary. The search committee doesn't want to hire someone who leaves after 3-5 years (the typical length of time given to a new recruit to obtain grant support)—this is a waste of time and money (i.e. the start-up package). Because the search committee is looking for a candidate who has an excellent chance of getting funded, a well-written description of your research interests is vital. It doesn't hurt to structure this a bit like a grant application with background, hypothesis, and specific aims that will be tested along with a brief description of the experimental approaches.

Leave out most of the technical details for two reasons. First, you want to focus more on the big picture than minor details. Second, you don't want to give away any secrets. Certainly explain how the planned studies are important and will greatly advance the field. The health significance of your research project is good to include if you are applying for a job at a medical school or otherwise expected to get money from health-related funding sources.

Including figures is generally a good idea. Pay attention to the attractiveness of the layout, just like in a grant application. There is often a page limit, or recommended length, and it is a good idea to come close to the upper limit on this—don't go way over the page limit but also don't turn in something much shorter than the stated page limit—this will be considered too superficial by the search committee.

References to key published articles are good to include, but don't overdo it and turn in a document with a bibliography that is longer than the research plan. Cite the key papers in the field for the background and relevant experimental techniques if not standard things (no need to reference things that everybody knows how to do).

Most importantly, give your draft to several colleagues for their input. Get advice on the overall directions of your proposed research as well as more mundane issues like typos and grammatical mistakes. Although this last point is rather obvious, not everyone seems to know it; I have seen numerous applications that are poorly written and these applicants are never seriously considered for an interview. One typo may be overlooked (or not, depending on the search committee) but multiple typos show sloppiness and this is not the image you want to project.

It is a good idea to write a general template of your research directions but then tailor it for each place so that you stress the relevance of your research to the specific requirements listed in the job description. This usually involves just minor changes in the emphasis of your plans and not fundamental changes in your overall research direction. By taking the time to customize your research plans, it will also show that you are serious about the position and not just applying to hundreds of openings. If you are actually applying to a large number of openings, you can prepare a few versions of your research plan that stress different things and then submit the most appropriate one to each place you apply.

Research Plan Template

My overall research focus is (insert name of department, or phrase stressed in job description), which has been (the major focus / one of several key directions / the hidden agenda) of my research for the past (10 years / year or two / twenty minutes). Specifically, I plan to use tools in the fields of (insert relevant words from job description, such as stem cell biology or nano-technology or whatever is currently hot) to study (insert the name of your research project). The central hypothesis is (insert something plausible but not 100% certain yet). I will test this hypothesis in the following three Specific Aims.

Aim 1: (describe the objective that is farthest along, and for which you have the most data)

Aim 2: (describe the aim that is the next farthest along)

Aim 3: (describe your wild and crazy aim that, if it works, will be a really big advance to the field)

(Expand on each of these aims in detail, including key figures/diagrams/tables to provide background information of models and preliminary data, if any.)

In summary, this project will greatly advance the field of (state name of department again) and provide a better understanding of processes that are (fundamental to science / relevant to disease / probably irrelevant but really cool).

The Cover Letter

You need to send in a cover letter with your CV and research plans. The purpose of the cover letter is to succinctly state your qualifications for the position and entice the reader to spend more time reading your application. The cover letter is very important and needs to be customized for the place to which you are applying. As with the description or your research plans, you can prepare a template and just change a few of the details. Typos and grammatical errors will have a negative impact and you need to proofread carefully. Ask a friend to look it over.

Standard approach

Dear Search Committee Members: (this is a default; if the job advertisement lists a name, use this instead)

I would like to apply for the position of _____ that was described in _____ (mention the source of the advertisement). I have attached my CV and a description of my research plans. In addition, I have (attached a list of references for you to contact / requested letters of recommendation to be sent on my behalf).[1]
(In the next paragraph or two, briefly describe the major findings of your thesis and post-doctoral projects. Then, summarize your future research plans in another paragraph. Stress the aspects of your past and future research that are most relevant to the requirements of the job.)
In conclusion, my qualifications precisely match those of the job opening listed in the advertisement, and I am looking forward to hearing from you.

Sincerely, (your name)

---

[1] Some places just want a list of references while others want you to arrange for letters to be sent. Make sure you follow the requested procedure or your application will not be seriously considered.

Non-standard high-risk approach that just may work (or not!)

Dear Future Colleagues:

Today is your lucky day. You can stop wasting your time reading CVs and just hire me. I am perfect for the position, as you can see from the attached CV and description of my future research directions. Because I am convinced that you will agree, I've already planned to move there and start work immediately. I have even bought a house and enrolled my kids in the local school system. See you next Tuesday.

Sincerely, (your name)

Chapter 3
The Job Interview

Most post-docs think they know how to interview—after all, they successfully interviewed at a graduate school and in many cases also had to interview for a post-doctoral position. However, the "real" job interview is not like the previous ones because you are not merely moving up—you will be moving out into something new. The transition from post-doc to assistant professor (or comparable position as lab director in research institutions) is bigger than the difference between college and graduate school. In college and in all your education up to that point you only had to show up to classes, listen, read some books, digest all the information, and grow. This was the "egg" stage of your career.

You did well as an egg and got into a good research laboratory doing real science! But real science is not like research in your college lab courses—those usually had a defined problem and one specific solution. In the real research environment the first step is to identify the problems and then go about finding the solution (or in some cases, finding the solution first and then searching for the problem you just solved). This is the "caterpillar" stage of your career—caterpillars need to move around the environment seeking out food, but are limited in their mobility. When working as a researcher in someone's lab, you are usually limited in how far you can go and what you can do.

If you succeed as a caterpillar and grow big, you can compete for an independent position heading your own laboratory. But most search committees don't just want a large caterpillar—they want something more than just an excellent researcher who can do the technical parts of science. Some of the additional tasks required of an independent scientist are listed in the table below and described later in this book. Most importantly, the science projects you can pursue in your lab are not limited by anything except your imagination, your ability to find collaborations to help with technical aspects beyond your capability, and your success in getting funding. When you are the leader of a laboratory, you will have much more responsibility and mobility than during your caterpillar phase—you will be a butterfly.

During the interview process, the most important thing to keep in mind is that you need to show the search committee your potential to move beyond the caterpillar stage. In simplest terms, show them your inner butterfly.

School/coursework (the egg stage)
- Absorbing/digesting information provided by teacher
- Passing exams

Research/benchwork (the caterpillar stage)
- Deciding what projects to work on
- Designing experiments
- Writing drafts of research papers
- Presenting your work to colleagues (in your department and at scientific meetings)

Group leader (the butterfly stage)
- Deciding on the overall research direction
- Setting up collaborations
- Hiring and training lab members
- Getting funding
- Budgeting money
- Helping your trainees write research papers
- Giving talks at scientific meetings
- Teaching (for academic positions)
- Serving on committees and other administrative tasks
- Reviewing manuscripts and grant applications
- Editing books and journals
- Organizing scientific symposia

Helpful Advice for the Interview—the Dos and Don'ts (and One or Two Maybes)

- Dress appropriately.

> - Try to look like a nicely dressed faculty member about to give a seminar. Gauge it by the standards at your university—what do invited seminar speakers usually wear?

> - Don't show up looking like you're about to get married, wearing a three piece suit or formal dress. Conversely, don't interview in gym shorts and a tee shirt.

> - Maybe try to wear something comfortable, or maybe not! While some advice books say to wear comfortable shoes, I interviewed in the most uncomfortable shoes I ever owned (and did OK, getting multiple job offers). These shoes were so incredibly painful that I didn't wear them again once the interviews were over. In hindsight these painful shoes served an important purpose—they didn't let me get so comfortable that I forgot I was on an interview. The throbbing pain of sore feet served as a reminder that I was there for a purpose.

- Look professional

> - If you don't already own an appropriate bag, now is a great time to get something for your computer and papers (your schedule, campus maps, reprints, extra copies of your CV in case some of the people you meet don't have it).

> - Don't take a backpack like a student unless that is the style of the faculty where you are interviewing. Often you can gauge this based on the place you are

currently working—what do the faculty haul their stuff around in? You want to do the same on your job interview, looking like a professor and not like a student.

- Take it seriously and plan things carefully but also relax and have fun.

- Don't get so nervous you forget something important in the hotel.

Check list before leaving hotel room

__ Professor-like bag with your computer, disk/drive containing lecture slides, and essential papers such as directions to the place you're going and your schedule.

__ Shirt, hopefully ironed. If you forgot to iron it, it's too late now so just keep your jacket on the whole day.

__ Jacket—unless you're sure the place won't expect it for the applicants (both men and women). If no jacket, remember to iron your shirt.

__ Socks/stockings and shoes. Albert Einstein was known to show up in shoes without socks, but this was after he had a job and was famous.

__ Pants (put on and zipped shut). As an alternative to pants, women (or very liberated men) can wear a skirt or dress if not too revealing.

- Try to bond with the faculty members of the department and show them you can be a good colleague who will be useful to have in the department. Collaborative research project ideas are a plus.

- Address them as "Professor," "Doctor," or "Larry / Susan" (if their names happen to be Larry / Susan and they give the impression that they expect you to use their first names).

- Avoid referring to them as "dude."

- Don't treat everyone as an old friend, especially the chairman of the department who you met for the first time at dinner last night.

- While it is important to have a sense of humor, you don't want to come across as non-serious or an intellectual lightweight. Goofy is bad. Witty and clever are good.

- Do your homework—learn the research interests of the faculty and think of questions in advance.

Good: Read a recent review or two that they have written and look at their webpage to learn about their research projects.

Bad: Read reviews that are critical of their theories and raise these points. Play devil's advocate and attack them on their work.

- Always ask the people you meet about their latest research. Even though you should have a general idea of the central research topics from the publications you read, those publications are what they did in the past and are not always the current directions.

- Don't cram so hard the night before your interview that even seventeen cups of coffee can't keep you awake.

- Do drink a little more coffee than normal—the caffeine buzz will help you get through a long day.

- Don't drink more coffee than you can handle without becoming excessively energetic. Remember—caffeine is a strong diuretic and you may not have time to hit the lavatory every 20 minutes during your interview and job seminar.

- Prepare several versions of the description of your previous research projects and future plans. Although you will give a seminar on your research, not everyone you meet during the day may be able to attend your seminar. You want to have several versions ready.

> - The long version—about 10 minutes. This is rarely needed and may only be appropriate for situations where the person you are meeting missed your seminar and wants you to give them the details.

> - The regular version—about 3-5 minutes. This is the ideal length for an interactive conversation. Expect questions that will allow you to expand on topics that the interviewer finds interesting.

> - The short version—about 1 minute. This is needed for casual meetings in the hallway with people who ask what you work on but don't want to stand there for 5-10 minutes while you ramble on.

Common and Not-so-common Interview Errors

Annoyingly repeat every question asked during the interview

> You (speaking to the professor, who is obviously irritated): "That's an excellent question. Why <u>do</u> I want to join the faculty at this institution? Well, let me tell you why I want to join the faculty at this institution. The reason I want to join the faculty at this institution is..."

During seminar, not making eye-contact with audience. Ever!

Assuming that "casual Fridays" applies to job applicants too!

Dressing up in a costume for your seminar, convinced that it will help the audience remember you.

The Job Seminar

Many applicants make the mistake of treating the job seminar like a regular seminar. It isn't. While there are certainly many similarities between a regular science talk and the seminar you have to give as part of the job interview, there are several major differences.

•   Give a good introduction to the topic.

> A well-planned introduction will show them that you will be a good teacher. This is especially important if the job will require a large amount of teaching. Normal seminars at scientific conferences are usually quite specialized and are geared at scientists in the field. For the seminar at your job interview, you want to make sure that a broad audience of faculty and students understands your research area and accomplishments. Avoid jargon.

- Be clear about what <u>you</u> have done.

    It's OK to say "I" instead of "we" when talking about your own work. The members of the search committee who are listening to your job talk will be interested in what you specifically have done versus what was done by your collaborators, if any. Collaborations are good—they show that you can work with others on a team. You don't need to pretend you did everything yourself. Give proper credit to others, but certainly emphasize what you have done yourself.

- Spend a few minutes on the future directions you want to pursue.

    Some universities have a separate hour (or two) in which the candidate presents his/her future directions. But even if this is the case, it is a good idea to briefly go over them at the end of your seminar.

- Be enthusiastic about your work.

    As obvious as this sounds, some job candidates present their work with the same enthusiasm that typically accompanies a trip to the dentist for root canal surgery. If you're not excited about what you have found, how can you expect your audience and especially the search committee to be excited?

- Have fun!

    This is also obvious but very important and commonly overlooked. If you are in agony suffering from stage fright it will be equally painful to your audience. Practice enough so that you are comfortable presenting your work to an audience but don't practice so much that you become bored by the subject (see above point).

- Leave time for questions.

  Show them you really know the field by your ability to answer questions and discuss things a little outside your main focus.

  Questions help you evaluate them. The interview is a two-way process; you are also interviewing them. If there are not any good questions after your seminar, they don't really understand your work. When starting a career, it is better to be at a place where the other members of the department understand your system, and even better, can raise intelligent questions. Leaving time for questions after your seminar will give you important information about the level of students, post-docs, and faculty at the place you are interviewing.

How to answer a difficult question

Great answer:
  That's an excellent question. The experiment has been done and I'll know the results when I get back to the lab. I can e-mail you the results if you would like to know the answer.

Good answer:
  That's an interesting question. Let me give it some thought and talk to you about it after the seminar.

Weak answer:
  Well, I thought of doing that but it was too much work.

Bad answer:
> That question makes my brain hurt. Can't you ask me something easier?

Terrible answer:
> That is the stupidest question I ever heard. In fact, that question is so stupid that I won't even attempt to answer it.

Worst possible answer:
> I thought you'd never ask! I assumed you were too dumb to realize the glaring error in my model.

The Informal 'Chalk Talk'

In addition to the regular seminar, many search committees for academic positions ask the candidate to present an outline of their proposed research plan. This is often called an informal chalk talk.

- Chalk does not always mean chalk.
  - Ask beforehand if you should prepare slides.

- Informal does NOT mean unimportant.
  - Prepare your chalk talk in advance and practice in front of friends who can critique your presentation.
  - Anticipate questions and have answers ready.
  - Show them you can think on your feet.

- Present a clear outline of your next 3-5 years.
  - Have a main project with 3 or 4 specific aims.
  - A second project is good.

- It's OK to expand into new areas/techniques.
  - Collaborations are an excellent way to move into new areas. Potential collaborations with faculty in the department and/or school are a plus.

- Have a general idea of where you want to be in 5-10 years.
  - What do you want to be known for as the world's expert?

Dinner

You've finished your seminar and interviews for the day and now you think you can relax over a nice quiet dinner. That is true only if you are dining alone or with friends/family who live in the area. If you are having dinner with the department head and/or other members of the department, you are still on the interview.

Some advice books actually suggest that you avoid drinking alcohol during the interview but this will not help you create the perception that you are socially well-adjusted if everyone else at the table has a glass or two of wine and you're the only one not drinking. If you never drink alcohol, don't start now—just order a non-alcoholic drink. But if you are comfortable drinking alcohol, then go ahead and join them if they drink, but avoid drinking more than anyone else. If they order one bottle of wine for all 6 people at the table, then sip slowly and make it last through the meal. If they each have a round at the bar while waiting for a table then join them in moderation but don't drink more than you know you can handle and still be coherent when talking science.

Be aware of current events—read the newspaper and know what's happening at the global, national, and even local level. It's not a bad idea to learn about local sports teams just in case it comes up in the conversation (rare among scientists but it can happen). Use the internet to find out about local cultural and recreational activities and if any of them interest you, bring this up ("I've heard that this city / town / village / hamlet has a very nice art gallery / opera hall / bowling alley / gas station").

Most importantly, don't forget your table manners. If you don't have any, now is an excellent time to learn. As obvious as this advice sounds, every so often a candidate seems to have been raised in the wilderness by a pack of wolves. If you have no idea which bread plate or water glass to use, and are totally overwhelmed by a napkin and multiple forks, go out to a fancy restaurant with some bipedal friends and practice before going on interviews.

Great topics for dinner conversation

- Science (the big picture—not just more details about your own research).

- What it is like to live in the area.

- Advantages and disadvantages[2] of working there.

OK topics for dinner conversations

- Sports, especially if the school has a strong college sports team.

- The problems with getting research funding.[3]

---

[2] Don't expect too much on this topic. But ask and if they launch into a 30 minute tirade about how awful it is to work there, hope that you get a job offer from another place.

[3] This is a chronic problem that may come up spontaneously in the conversation but which will not make anyone very happy. You want the dinner to be fun, not a depressing event.

Bad topics for dinner conversations

   - Politics (unless they bring it up).

   - Religion.

   - Off-color jokes.

   - Your medical history.

   - Your prison record.

   - Why you're so desperate for a job you'll even consider joining their pathetic institution.

Waiting for the Job Offer

After you've gone on one or more interviews, you have to sit back and wait. You think you did well, but there were some things you're not sure about. You keep replaying these in the tape loop in your head. Maybe I should have said ____ instead of ____! In reality, it probably doesn't matter at all. They will decide based on many factors. Hopefully. Or maybe not! According to Malcolm Gladwell, author of Blink, we make critical decisions in the first couple of seconds.[4] If this is true, nothing you said matters—it's how you were perceived when you entered the seminar hall or interview room. In my experience the first couple of seconds is not the most critical time for an interview—it is the last 5 minutes that matters most, once the standard things are out of the way. Is the candidate able to think about totally new

---

[4] Blink: The Power of Thinking Without Thinking by Malcolm Gladwell (2007).

things? Can he/she come up with bright insights to problems a little bit outside his/her main focus?

Once the interview is over, you are left with weeks or months of waiting for the letter—hopefully the letter with a job offer, and not a rejection letter. During the wait, you can write to the school(s) with an update. This is a good idea if there are major developments.

<u>Definitely write to let them know if</u>

Your submitted papers have been accepted for publication in a good journal.

A grant application of yours was funded.

You received a major prize or award.

Another place offered you a job, and needs you to give them an answer in the near future.

<u>Don't bother writing them about minor things</u>

A paper listed as "in press" has come out in print.

Your submitted paper has been accepted by a journal of last resort.

_____

Multiple Offers—How to Choose?

If you are lucky, you'll get multiple job offers that you can then use for negotiation and obtain an even better offer. But be careful—some places will simply offer their best deal up front and if you start to haggle, they may not take kindly to the idea of bargaining. It is a delicate situation that you have to feel out and play your cards appropriately. If you really don't want to go to a particular place, don't string them along and use them to get a better offer somewhere else. You have to be honest and keep in mind that you are going to be joining one of these groups. On the other hand, if you are considered to be in demand it will raise your status and help secure your best offer, so don't be shy in sharing information on your job offers with the various places.

Negotiating the best offer and deciding which one to accept often involves many factors. While the reputation of the institution and size of the start-up package are important, also consider the geographic area—would you be happy living there for the rest of your life, or at least for 5-10 years? If you haven't consulted your spouse yet, now would be an excellent time to do so.

The main focus should be on the place and how it will help your career in the long run. It is great to go to a big-name place where there are stars, although most students will want to work in the labs of the famous professors and not with an unknown young assistant professor. Also, places with lots of big names can also mean lots of big egos which can get in the way of collaborations. Is there evidence of interactions among laboratories in the places you are considering? You want a supportive environment aimed at fostering collaborations.

It is good to consider who is in the department, as in the following example.[5]

---

[5] With apologies to Bud Abbott and Lou Costello for "borrowing" the idea from their famous "who's on first" sketch.

Who's in Your Department?

The following is a conversation overheard at a large annual meeting between Dr. Smith and Dr. Jones. Apparently, Dr. Jones had just taken a new job at a large university.

Dr. Smith: Hey—Jones—how are you? I haven't seen you since you started your new job.

Dr. Jones: I'm doing well. I joined a great department with lots of good people.

Dr. Smith: Do I know anyone there?

Dr. Jones: Well, there are many famous people in our department. Do you know Hu?

Dr. Smith: No I don't know who?

Dr. Jones: Well, he's very famous.

Dr. Smith: Who is?

Dr. Jones: That's right.

Dr. Smith: What's right?

Dr. Jones: No, Watt's in Cell Biology.

Dr. Smith: Who's in Cell Biology?

Dr. Jones: No, Hu's in our department.

Dr. Smith: How should I know who? You're not telling me! Do you have a chairman?

Dr. Jones: Nohbody's our chair.

Dr. Smith: You got nobody for chair?

Dr. Jones: That's right.

Dr. Smith: Then are you looking for a chair?

Dr. Jones: Why would we? Nohbody's good enough to be chair.

Dr. Smith: I agree. But please tell me the names of your colleagues!

Dr. Jones: We have two Kalleagues—Joe and Ann. They're married.

Dr. Smith: (getting frustrated) What's the last name of Joe and Ann?

Dr. Jones: No—I already told you Watt's the last name of a professor in Cell Biology!

Dr. Smith: Who's in Cell Biology?

Dr. Jones: No—Hu's in our department!

Dr. Smith: How should I know who? Look—would you *please* tell me the names of the people in your department!

Dr. Jones: Well, in addition to the people I've been telling you about, we got Problims.

Dr. Smith: It figures that you would have problems.

Dr. Jones: Sounds like you know him well.

Dr. Smith: Who?

Dr. Jones: Not Hu—we were talking about Problems.

Dr. Smith: I really don't understand. I just want to know the names of the people in your department. Could you *please* tell me?

Dr. Jones: That's what I've been trying to do. In my department Nohbody's doing what needs to be done, our Kalleagues are always working on grant applications, but Hu isn't. Not a day goes by without encountering Problems. We also have Disorder, Chaos, Mayhem, Procrastination, Backstabbing, Lethargy, Ignorance, and Benign-Neglect.

Dr. Smith: Sounds like you have troubles.

Dr. Jones: Not in my department! The Dean's office has Trubels.

Dr. Smith: I can imagine they'd have troubles.

Dr. Jones: So you know him?

Dr. Smith: Who?

Dr. Jones: Not Hu, Trubels! Hu is in *my* department!

(At this point, the conversation ends abruptly when Dr. Smith chokes Dr. Jones to death, leaves his lifeless body in a heap, and runs screaming from the room.)

It's not just who's in your department that matters...on occasion it also matters who's in your local pub!

Cambridge, 1953. Shortly before discovering the structure of DNA, Watson and Crick, depressed by their lack of progress, visit the local pub.

Chapter 4
Starting Up a Laboratory

Congratulations on your new job. You worked hard to get to this position, spending long hours as a student and post-doc, and you came out ahead of all the other applicants for the position. That is worth celebrating! But now it gets even harder, so don't celebrate too much. You have so much to do that it can seem overwhelming. There are some obvious priorities, such as setting up your laboratory—you can't write papers until you have data, and you can't get data until you set up your lab.

A typical list of priorities for a new investigator
1) Set up lab

2) Do experiments

3) Write papers

4) Write grant applications

5) Rewrite and resubmit papers and grant applications that were rejected

6) Write even more papers and grant applications

7) Repeat steps 2-6 until grant applications are funded and papers are accepted

8) Everything else (teach, serve on committees, etc.)

All you have to do is repeat steps 2-8 over and over and over, for the next 30-40 years. Some of these topics are the focus of other chapters in this book; the present chapter covers the various aspects of setting up your lab.

Lab Renovations

If you have special requirements for your research lab or are moving into a very old space that hasn't been renovated for years, the department may offer to update the space before you move in. It's good to check if the costs of the renovation will be in addition to your start-up package or if they will be taken from the money you were promised. If the latter, be aware that renovations can be very expensive. Renovations can also take much longer than anticipated, so it's best to start well in advance. Don't wait until you arrive and then discover that you need to find temporary space for months until your lab is ready. This happened to me—even though I started planning my lab many months in advance, the renovations weren't done by the time I arrived and I was given a temporary desk in a dark and dingy equipment room for a month, right next to an old incubator for bacteria that made a loud racket, making it impossible to talk on the phone or to concentrate.

Planning a new lab can be difficult. After all, we were trained in science, not in architecture or design. It's hard to know what layout is best for your new lab—where you want the sink, the lab benches, the desks, and how many drawer units you need versus cabinets. In hindsight it is easy to know what you should have done, and the hard part is trying to guess your future needs. If you are doing the same type of experiments that you did during your previous research, you can base your design on that of the previous lab. Also, talk to people who recently set up their own labs and ask them for ideas.

To make your lab appeal to students and post-docs, consider a separate area or room where lab members can eat breakfast, lunch, and dinner, and drink coffee while analyzing data. For extra student-appeal as well as greater productivity, consider adding an automatic cappuccino or espresso maker and small fridge to hold lunches and snacks so they can work longer hours. Although popular with students, avoid installing a hot tub, a large flat-screen high-definition TV, or a wine bar.

Having the right equipment helps you succeed!

1865: Kekulé, moments before his brilliant insight
into the structure of benzene.

Buying the Right Stuff

During your previous training, you probably didn't have to make a large number of decisions about what to buy to equip a lab. If you joined a lab that was already established, you just needed additional supplies for your specific project. It's a lot different setting up a lab from the beginning. How do you know exactly what you're going to need? If you are going to study the same general things that you worked on during your previous training, you can simply buy all of the same equipment and supplies. But if you copy the complete inventory of your past lab, you will end up with much more stuff than you need—most labs have shelves of equipment and chemicals that are no longer used. Assuming you have a specific cap on your spending, you want to spend the money wisely so that you can maximize your productivity.

If your start-up package has a very short time window to use, then you have to decide on everything very quickly. It is better if you have a year or more to spend the start-up funds—then you can buy the absolute essentials that you'll use in the first few months and additional things as you need them. Often, you can find old equipment around your department that is not being used, and the owner will be grateful if you take it off his/her hands either as a gift or a loan. A limited budget can also be stretched by using equipment in other labs. Do you really need your own spectrophotometer if you plan to use it for an hour once a month? If there is equipment in your department that is rarely used, it may be possible for you to avoid purchasing your own. When I started my lab, I met with another new faculty member and we decided to share a number of expensive pieces of equipment—some I bought and housed in my lab, the others she bought and housed in her lab. We avoided duplication and allowed our start-up packages to go much further.

Most places have administrative staff to help you handle the budget and assist with ordering. Otherwise, if you have to do this yourself it can be a painful process that seems very far removed from science. Computer programs can help but are generally not geared for the specific task of running a scientific lab. Wouldn't it be great if there were computer programs for professors?

Product Review: Computer Programs for Professors

The makers of the popular program AssistantProfessor have come out with a new release, version 2.0. While this is a large improvement over the previous version, the new release still suffers some serious problems. Several other programs are not compatible with AssistantProfessor 2.0, including popular programs such as CollegeBuddies (all versions), Wife (1.0 or 2.0), Children (1.0, 2.0, or 2.1), or any release of LifeOutsideLab. Even when running as the only program on your system, AssistantProfessor 2.0 will often crash and start all over at the beginning.

Like the earlier version of this program, AssistantProfessor 2.0 will terminate after 5 years (6 in some locations) unless you upgrade to AssociateProfessor. There are two versions of AssociateProfessor: tenured and non-tenured. The deluxe tenure edition is extremely hard to get since it comes with guaranteed lifetime support, although there's a good chance the software will become obsolete in 5-10 years unless it's upgraded to FullProfessor. The non-tenure version of AssociateProfessor is easier to install but it expires every year and needs to be reinstalled. Also, the non-tenure version cannot be upgraded to FullProfessor in most locations. If you do upgrade to FullProfessor you will get a free add-on program, DeadWood, although this program doesn't seem to do anything useful and takes up an enormous amount of space.

Despite all the problems, AssistantProfessor 2.0 and related programs are very much in demand, with hundreds of people vying to obtain each new copy released to the market, even if available only in some remote location.

Setting Up the Lab—Hiring the Right People

One of the hardest tasks for someone just starting out is to find the best people to hire. Depending on your start-up package, you may have money to hire only 1-2 people, so it is critical for your career that you hire the right people. If you make a mistake and hire somebody who can't do experiments, or who you don't get along with, it will be a disaster. Typically, a new investigator has a limited start-up package of 2-3 years, and in this time you need to get results that you can publish and use in grant applications. To complicate matters, most of the highly competent students or post-docs don't want to work for an unknown young assistant professor, and would rather work for a famous scientist with a solid reputation for training people. So, you have little chance of finding anyone as competent as yourself. You'll have to settle for the best you can get and learn how to be a good coach, manager, and cheer leader to motivate them. Most important is the selection process. Sometimes you get an applicant who seems too good to be true; often they are. Always get references, preferably from someone you trust to give you an honest opinion. Make a phone call to speak with one or more of the references; don't rely on the written letter alone because these rarely contain anything negative.

You will likely be faced with a difficult choice—do you select someone with extensive experience or opt for the neophyte with little or no experience? There are potential advantages and disadvantages you should consider.

<u>The seasoned veteran—many years of experience working in a number of labs</u>

Plus: won't need years of training

Negatives: may not want to learn your way of doing things and instead do things the way he/she learned in the past.

Summary: could be great, but check references carefully—call the past mentor and ask about potential problems. Rarely will a past mentor be blunt and say the person is terrible. More often, you need to interpret his/her comments.

Interpreting Recommendations

When the previous mentor says:
    We both felt it was time for him/her to move on after one year in my lab.

It usually means:
    Personality clash! Time to move on would be a valid reason if it really had been a long time (three years or more in most fields, although on occasion shorter training periods are appropriate).

_____

When the previous mentor says:
    He/she is very independent.

It usually means:
    He/she is not going to listen to a youngster like you, and will just do whatever he/she wants.

_____

When the previous mentor says:
    I really wanted to keep him/her, but ran out of funding.

It usually means:
    The person was not the star of the lab, especially if the mentor kept everyone else (and hired even more people)—if so, the "lack of funding" was just an excuse to get rid of the weakest link.

The budding young scientist—no previous experience

Plus: usually very energetic and enthusiastic, hasn't learned the wrong way of doing things.

Negatives: no reference letters from scientists who can evaluate his/her ability to work in a research laboratory. Recommendation letters from college teachers saying that he/she showed up in class, asked questions, and did well on exams doesn't count for much. The person will probably need extensive training. If the job candidate is thinking of applying to medical school and just wants research experience to make the application look better, you'll lose him/her after 1-2 years, and you'll also have to send out dozens of recommendation letters.

Summary: this can be a tough decision, and much depends on your gut feeling after the interview. Even then, there are people who interview well but aren't so good in lab and can't use their knowledge to solve problems. Consider giving a quiz to assess the applicant's general problem-solving skills, with questions appropriate for the level of training they should have received and which are relevant for your research. For example, anyone who took a chemistry course in college should be able to answer "how would you make 100 ml of 1 M sodium chloride?"

Great answer: Weigh out 5.8 g of NaCl, add to about 90 ml of water, dissolve, and then add water to exactly 100 ml using a graduated cylinder.

Good answer: Add 5.8 g of NaCl to 100 ml of water.[6]

Bad answer: Well, I know how to make 1 liter of a 1 M solution—add 58 g of NaCl to 1 L of water. Then, after dissolving, measure 100 ml and discard the rest.[7]

Terrible answer: Isn't there a kit to do this?

Worst possible answer: I would come to you for help. (This would be a great answer to a complex problem, but you don't want to be bothered for every little thing.)

The above example requires background knowledge, and isn't simply a test of problem-solving skills. An alternative approach is to ask questions that don't require previous knowledge, and which test the candidate's ability to solve new problems. A popular book on the hiring process at Microsoft describes a number of mental puzzles—things most people don't know the answer to (unless they are into nerdy puzzles) and which supposedly test the ability to solve abstract problems.[8] But be careful with this approach—the standard answer may not make the best scientist. For example, consider what would happen if you interviewed a highly qualified job applicant with years of lab experience, and asked him/her a typical puzzle used by Microsoft to evaluate job candidates.

---

[6] This does not take into account the volume occupied by the 5.8 g of NaCl, and therefore is not the best answer, but it is an acceptable answer because you only specified 1 M, and not 1.00 M NaCl.
[7] I used this quiz when first hiring people, and one applicant said this as a serious attempt at an answer.
[8] How Would You Move Mount Fuji? Microsoft's Cult of the Puzzle— How the World's Smartest Company Selects the Most Creative Thinkers by William Poundstone (2003).

If Scientists Used the Interview Methods of Microsoft

Interviewer: Let me ask you a hypothetical question. You have a book of matches and a long slow-burning fuse, the kind used to light a firecracker. The fuse will burn for exactly 1 hour. But the fuse isn't even—parts of it will burn faster than others, and all you know is that the entire fuse will take 1 hour to burn. How can you tell when 30 minutes are up?

Candidate: I would look at my laboratory timer.

Interviewer: But you don't have a timer.

Candidate: But why do you want to know so precisely when 30 minutes are up if you don't even have a proper timer?

Interviewer: It's just a theoretical question.

Candidate: OK, then I would look at the theoretical clock on the theoretical wall.

Interviewer: There is no wall clock.

Candidate: How about the clock on the computer?

Interviewer: There is no computer.

Candidate: What kind of lab has no timer, clock, or computer?

Interviewer: You're in a remote location.

Candidate: Not even a laptop? Then I would look at the clock on my cell phone.

Interviewer:	You don't have a cell phone.
Candidate:	Sure I do. It's right here in my pocket.
Interviewer:	You can't use your cell phone.
Candidate:	You mean this job won't allow me to carry a cell phone?
Interviewer:	Yes—you can carry a cell phone on the job. You just can't use it to solve this theoretical problem. You have an hour-long fuse and matches and want to know when 30 minutes are up.
Candidate:	(looking perplexed) I have a cell phone with a clock in my pocket and I can't use it to let you know when 30 minutes are up?
Interviewer:	You are in a very remote location. There is no cell phone service. Your cell phone battery is dead. All you have is the hour-long fuse and matches.
Candidate:	Do I have a sundial?
Interviewer:	(getting frustrated) All you have is the fuse and matches! Look, it's really pretty simple. Just light the fuse at both ends—then when it burns out, you would know when 30 minutes are up!
Candidate:	How about if go on eBay and trade the fuse and matches for a proper timer?

Idea to Get Dedicated Workers—Start a Cult Religion!

Want to have dozens of energetic and motivated lab workers, but don't have money to pay them? Try this approach—put your contact information on the following advertisement and post around your community. Then, just sit back and wait for glassy-eyed disciples to come knocking at your door.

Are you feeling like your life has no meaning?

Are you tired of unfulfilled promises from your priest, pastor, rector, vicar, rabbi, guru, or imam?

Are you feeling spiritual emptiness?

Join the Church of Bioscientology!

We are deeply rooted in the principles of biology and science.

We will teach you fundamental concepts of science, and let you experience the thrill of discovering a new scientific truth.

We will help you achieve inner peace through pipetting and enlightenment through data analysis.

There is no affiliation between The Church of Scientology and our program on Bioscientology. We are completely independent. Any resemblance is purely coincidental.

The Other End of Hiring—Letting People Go

The best people in your lab will naturally want to move on when the time is appropriate. You would obviously like to keep these people as long as possible, but being a good mentor means that you need to respect their efforts to climb the ladder and get to the point where they head their own labs. The middle level people in your lab may not want to move on as quickly, and often need to be prodded into action. For this group, you don't mind if they stay an extra year because they are moderately productive, and while not in the top group, they certainly pull their weight. For them, a simple deadline is often enough. The problem is the bottom group, those who are not contributing to the lab at all. If you are lucky and/or good in hiring, you can avoid this group entirely. I haven't been so lucky and have taken on some people who seem to be completely different than the people described in the letters of recommendation.

Firing is not easy to do well. In my experience, the best way is to meet with them, have an honest discussion about their career goals, and gently point out that they aren't fitting in with the laboratory. In most cases, the feeling is mutual, and the about-to-be-fired person is not happy in the lab and would rather do something else. The conversation then becomes "how can I help you move into the new direction" rather than "you're fired." But in some cases, the person is unaware of the problem and thinks that he/she is doing as well as everybody else in lab and doesn't want to leave. It may help to give the person an opportunity to put more effort into the project (but with a deadline and another meeting to discuss the progress). The bottom line is to be gentle, not just because this is a human being you're dealing with, but because he/she still has keys to your lab and can inflict serious damage if upset. (I've heard horror stories of fired employees exacting revenge on the lab!) The following are examples of what you want to say, but what you really should say instead.

What you *want* to say: What the #%!* do you think about all day?

What you *should* say: You seem to be distracted. Is there something on your mind?

_____

What you *want* to say: How did you ever get a degree? Did you bribe someone?

What you *should* say: I'm not sure that your background has adequately prepared you for the challenges of this project.

_____

What you *want* to say: You have no future in science. You can't think your way out of a paper bag!

What you *should* say: You may do very well in a different field of science...political science!

_____

Chapter 5
Running a Laboratory

While the focus of a scientist is science, this does not always mean benchwork. When running your own lab, you have to learn to be an effective mentor to your students, post-docs, technicians, and other trainees that work with you. This chapter contains general advice (and random attempts at humor) for some of the non-science aspects of running a lab.

The Laboratory Interactome

When you start out, it is likely that you will have just a few people in your lab, and that you will interact with all of them on a frequent basis. The following diagram illustrates this simple system.

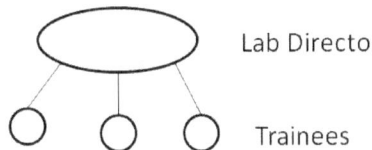

Lab Director

Trainees

Interactome of a small research group. Each of the trainees (students, post-doctoral fellows, technicians, or others) reports directly to the lab director.

This system is fine for a small lab. Although your students and post-docs are not getting any experience as mentors with this structure, chances are that they won't mind. They probably will welcome the opportunity to focus on their own scientific projects without the distractions that come when a junior person is asking lots of questions.

If your lab grows in size, it is difficult to find the time to meet with everyone on a daily or even weekly basis. Some labs just keep the original model for interactions and scale up, creating the following interactome.

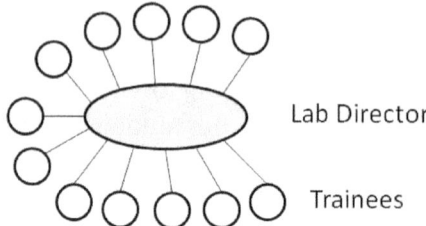

Lab Director

Trainees

Interactome of a large research group, based on the small group system in which each trainee reports directly to the lab director.

While it could be argued that this model is best because all trainees are being supervised by the most highly qualified person (i.e. the lab director), there are usually problems. First, the director is often bothered by simple things that could easily be solved by others in the lab, and he/she is probably feeling overwhelmed with running the lab. Second, the trainees are not getting any experience as mentors, which is important later in life.

Instead of the simple model that worked for your small lab, it is best to structure your lab such that new people have one or more people that oversee their projects and provide daily interactions. The following scheme is an example.

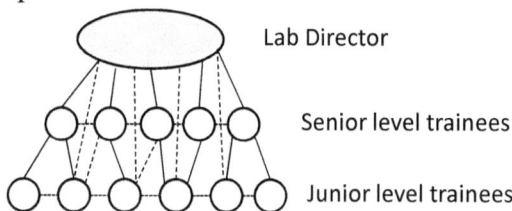

Lab Director

Senior level trainees

Junior level trainees

Interactome of a large research group arranged into tiers in which the lab director interacts most frequently (solid lines) with a subset consisting of the senior trainees, and less frequently (dashed lines) with junior members. The junior members interact more frequently with more senior members than with the lab director. This provides mentoring experience to the senior lab members and allows the director to concentrate on the important questions. The various lab members also can interact on joint projects (dotted lines).

For such a model to work, it is important to set it up when new people start rather than just hope that the junior person seeks out the right person to ask when they have a problem or questions. The lab director should include the senior trainee when meeting with the junior trainee. Over time, the junior person will need less supervision from the senior person, eventually becoming an independent senior trainee able to supervise newer lab members.

Shared research projects can be a valuable way to boost productivity and allow for a lab to tackle a larger project. To work most efficiently, this requires direct interactions among the project members. Other peer-to-peer interactions include training; the best way to learn something is to teach someone else. Rather than the lab director teaching each person separately, he/she can train one or two people and then have them pass the training on (occasionally checking to make sure the training is done correctly).

If collaborating with other laboratories, instead of having just the lab directors exchange ideas, it is better to have trainees interact directly. This can involve physically travelling from one lab to another, or interactions through e-mail and other communication methods. Such a scheme is ideal for solving complex problems.

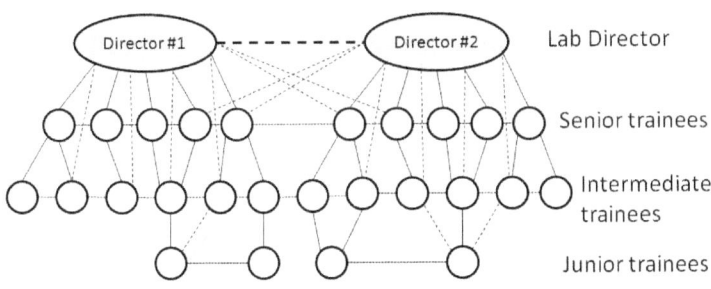

Complex interactome with multiple lab directors.

During my training, I spent 3.5 years in one lab as a graduate student and 3 years in another lab as a post-doc. Both of these lab groups were large, containing 20-25 people (students/post-docs/technicians). One lab was run using the pyramid system model and the average student graduated with a Ph.D. in 4 years or less. During my time there, the longest time a student took to get a Ph.D. degree was about 5 years. In this environment, it would have been very hard for someone to either slack off or get side-tracked on a dead-end project. The other lab I trained in did not use a pyramid system, but instead relied on the single interaction model. The motivated people in the lab voluntarily formed interactomes to increase their efficiency and provide advice to each other. But there was no formal system in which new members of the lab were supervised by senior lab members. When senior lab members noticed students slacking off, there was no simple way to bring these students up to speed. In this lab, there were three students I overlapped with who took 9 years to complete their Ph.D.s! It didn't help that the advisor traveled frequently and would spend weeks out of town. But even with a hands-on advisor, it would be difficult to be efficient with such a structure.

To encourage an interactive pyramid-like structure, when new people enter your laboratory, make it clear who they should consult with. Also, check that the senior people are taking the responsibilities seriously by periodically asking how the new people are doing. This will give the senior people valuable training as a mentor, although it will take time away from their own projects. It also helps to encourage sideways interactions. When my trainees ask my opinion, if I think someone in lab has an answer to this, I ask if this person was consulted—the idea is to encourage them to interact with each other.

Resolving Conflicts

The upside of a lab that is highly interactive is increased efficiency and decreased chance of outright fraud—usually not a problem, but there are many horror stories of students/post-docs/lab directors who created data to make a convincing story. A lab with multiple levels of interaction is less likely to provide an opportunity for exaggerated data, or outright fraudulent data to occur. The major drawback to this lab structure is that conflicts may arise. There are ways to minimize the chance of conflicts. From the beginning of a project that is shared by several people, be clear as to who is the overall leader of the project (other than the lab director)—this leader is the likely person to be the first author on the paper.

Even with the best plans, the uncertainty of science can derail these plans and cause conflicts when a contributor to a project ends up doing more than the main person assigned to the project. Or, sometimes to achieve a higher-impact paper, it is necessary to combine two or more projects. A common solution is to have multiple "first" authors on a paper. Some journals allow for even more than two co-first authors. In one case, I published a paper with four authors, and all four were co-first authors! Even with this solution, one name has to appear first on the actual paper. Many on-line databases (such as PubMed) do not include the asterisk indicating shared first authorship. If one person is not clearly best to put in the lead author position, consider drawing names from a hat (in the presence of the authors)—this avoids all chance of an appearance of bias. Or, make a deal—X will be the first author on this manuscript, but Y will be the first author on the second manuscript which is already in progress. Being fair is important—if your trainees feel that they are not being given appropriate credit, they will be less motivated to continue working, and may file a formal complaint which is not good for all parties involved. Most institutions have an office that deals with conflicts arising between mentors and trainees, and it is best to resolve conflicts before they get so far that outside arbitration is needed.

In addition to the conflicts that arise from choosing the first author, or co-first authors, there can also be problems when lab members think they deserve co-authorship because they contributed advice or a small technical part of one experiment included in the paper. Many journals state that a criterion for inclusion as a co-author is an intellectual contribution, and that merely providing technical assistance is not enough. This policy is also intended to discourage department chairs, or other senior members of a research group, to insist on authorship because the research was done in their group. Funding is a trickier subject—if a senior person was responsible for getting the money to support the people doing the experiments, does he/she qualify for co-authorship? And what about people who supply an important reagent—does this qualify for co-authorship? In my experience, I solved potential conflicts by asking people if they wanted to be a co-author and help write the paper, with the expectation that if they didn't want to help write and/or edit the manuscript, that they should not be co-authors.

Other conflicts can arise from personality clashes. If caught early enough, before the dislike turns to mortal hatred, it may be possible to reduce the interactions between these people by changing the projects each is working on. A larger problem is when the personality clash is between you and one of your trainees. If your institution has an office for conflict resolution, it may be helpful to bring up the problem with the appropriate person and try to find a solution.

A colleague involved with her school's office for faculty development offers this advice on avoiding potential conflicts between you and your trainees.

- Be friendly, but don't be a friend. Keep the relationship at the professional level—you are the boss. While it is good to get to know them on a personal level, don't become buddies with them. It is fine to occasionally take your lab members out for dinner or drinks to celebrate an important event (paper getting accepted, grant application getting funded, a trainee winning an award or graduating, or any milestone in a career). Avoid romantic relationships with anyone under your leadership.

- Treat your trainees with respect and fairness. Don't have your trainees compete against each other in order to move a project faster. If a trainee doesn't appear to be working hard enough, talk to him/her before you take the project away and give it to someone else.

- Ask your trainees about their career goals and expectations. Meet them on a regular basis—it is good to pass through the lab to talk, but also meet them in your office where they can talk candidly about their career development and any scientific problems that have arisen.

Being a Good Mentor

A mentor needs to provide guidance and training to students, post-docs, and technicians so that they can proceed most efficiently. But efficiency in the short-term does not always amount to long-term efficiency. Keep in mind that your objective as a mentor is training. It is easy to look at data obtained by your trainees and then tell them your ideas for the next few experiments. It takes more time to ask them what they would do next, rather than tell them what you would do. Sometimes they will come up with interesting ideas that you hadn't considered. If they have no ideas, rather than tell them yours, try to ask them questions that coax the idea out of them. This serves several purposes. First, it is a useful training exercise that will teach them how to think about the problem. Second, once they get into the habit of coming up with ideas, they will do this before they meet with you to show you the data, and so the process will become more efficient. Third, they will be much more motivated to work on a project if they come up with the ideas themselves (even if you dragged the idea out of them).

In addition to asking your trainees to interpret their data and come up with ideas for the next experiments, you can also ask your trainees to do literature searches to see how their results fit into the big picture. Rather than provide them all of the papers they need to read for a project, ask them to do the literature searches and see what they come up with. Maybe they'll find something you missed, and unless they fail to find anything at all, they will benefit from the sense of independence and control over their project.

You will need to give them feedback from time to time. Try to keep it positive; rather than say "that is the stupidest idea I ever heard" it's better to say "that is a pretty stupid idea, although not quite the stupidest thing I ever heard." Or, say something even less negative such as "I don't understand the logic behind that idea."

The following are examples of incorrect things for a mentor to say to a trainee. All of these are paraphrased from examples sworn to be true and/or overheard directly.

"This data should not have been recorded on lab notebook paper—data of this type would be more appropriate for toilet paper!"
> - said to a graduate student by a famous mentor.

"Did you ever read a scientific paper?"
> - said to the same student by the same mentor.

"You are short, female, and foreign-born—what are you going to do about it?"
> - said to a post-doc by a well-known mentor (who was a tall US-born male).

"That's not how we do science—that would be playing games."
> - said to a visiting scientist by the same mentor, in response to a solution to a problem. The solution later proved to be the key step in moving the project forward, which the mentor took credit for.

"A female scientist with children will never succeed at a first-rate university."
> - said by a famous department chair to a young assistant professor starting her family, who did manage to succeed at a first-rate university.

"You can always find another wife."
> - said by the same famous department chair to his post-doc, after the post-doc explained that his wife would divorce him if he didn't take time off for a vacation.

Finding the Inner Potential of Your Trainees

Everyone is good at something and weak at other things. Part of being a mentor is to identify the strengths and weaknesses of trainees. If the weaknesses are in things that are vital to a career as an independent scientist (such as writing, public speaking, and networking), you need to help improve these areas. But some things that are weaknesses at one level of a scientific career are not really weaknesses at the next. For example, a student who is not good at benchwork may make an excellent professor if he/she is good at seeing the big picture, mentoring and teaching students, writing, and general networking.

As a mentor you want to help your trainees with their weaknesses but you also have to consider the efficiency of your lab. While training is certainly important, so is the overall productivity of your group. It is usually most efficient to have people work on what they are best at, and not spend time on things they do badly. Therefore, it is a balance between being productive by focusing on the strengths of your trainees versus spending time trying to help them develop in new directions.

Writing Recommendation Letters for Other People

As a mentor, you want your trainees to do well and move on to great careers as scientists. It can also impact on your career—some universities consider your mentoring record when evaluating you for promotion. Have your trainees gone on to good positions in science-related fields, or dropped out and quit science entirely? For your first promotion, it will probably be too early to tell if any of your trainees have succeeded, but by your later promotion(s) this will likely be an issue. Therefore, you want your people to do well.

Keep in mind that when people read recommendation letters, they expect a bit of inflation—everyone is above average! Good people are great, great people are excellent, and excellent people are outstanding. I was once on a promotion committee when someone commented on a particular letter written by a senior scientist who did not feel it appropriate to exaggerate. The recommendation letter used the word "good" many times, but never used words like "outstanding" or "exceptional" to describe the junior candidate. The committee member evaluating the promotion of the junior scientist said "Dr. ____ must not be good because the recommendation letter said they were good." In this case, the senior letter writer sincerely meant that the junior person was good and deserving of promotion. Fortunately for the junior person, enough people on the committee knew the senior letter writer and what he meant by "good" and so the promotion went through.

Phrases to include in a letter meant to be positive:

-Outstanding ability/productivity

-Excellent laboratory skills

-Remarkable talent

-Able to ask critical questions and see the big picture

Negatives are bad and should be avoided (unless you intend to write a weak letter). Words and phrases to avoid in a letter meant to be positive:

-Although…

-Despite…

-Brain-dead pond scum...

However, you have an ethical obligation to point out serious problems, but in a nice, complimentary way.

Examples:

> Problem: The person never showered or did their laundry, and nobody else in lab could tolerate being within 10 feet of them.

> What you write: He was so enthusiastic about the scientific project that he would forego social activities in order to devote every waking hour to the project.

> Problem: The person was difficult to get along with and was on speaking terms with only a couple of people in the lab.

> What you write: She got along with some of the people in the lab.

> Problem: The person didn't listen to your advice.

> What you write: He was extremely independent and enjoyed solving problems entirely on his own.

> Problem: The person never bothered to learn English.

> What you write: She was so focused on research that communication skills were not a priority.

If you really don't want to recommend someone, either refuse to write the letter, or write a letter that is lukewarm but sounds upbeat. For example, if your goal is to write a weak letter, refer to them as Dr. if they have a doctoral degree, or Mr./Ms. if otherwise. It seems respectful, but if you spent several years training someone in your lab and you still call them Dr/Mr/Ms, there is usually a problem.

<u>Example of a weak letter</u>

Dear Sir/Madam:[9]

I am writing to recommend Dr. X for (a position in your laboratory / a job in your institution). I have known Dr. X since he/she started in my laboratory and began working on … (describe the project in great detail, spending most of the letter on a generic description of their results without commenting on the person's abilities).

Dr. X's research findings have been published in (name of journal), which is universally considered to be a journal of scientific research in the field of (re-state name of journal). This publication was, and still is, a publication in a scientific journal.

Dr. X was without a doubt a member of my (laboratory / department / university / species). I recommend him/her for the position he/she has applied for. I can say without hesitation and with utmost certainty that if you offer him/her a position, he/she will definitely have a job offer.

Please feel free to contact me about this candidate.[10]
Sincerely, (your name)

---

[9] By not taking the time to type in the actual name of the addressee, you imply a lack of enthusiasm for the candidate.

[10] This is generally a hint that you have more to say that you don't want to put into writing. Your phone number is on the letterhead (or should be indicated below your name if not on the letterhead) and they will contact you if they want more advice. Stating that they can contact you usually means "please call me—I have more to add."

Example of a strong letter

Professor/Doctor (insert appropriate name[11])
Chair, Search Committee
Department of Important Stuff
Big Name University
Archaic Snail Mail Address Line 1
Archaic Snail Mail Address Line 2

Dear (insert appropriate name):
I am writing to enthusiastically recommend Dr. Joan B. Goode for a position in your department. I have known Joan since (date) when (we met at a scientific conference / I was asked to serve on her thesis committee/etc). I was so impressed with her knowledge of a complex system that I offered her a position in my laboratory, which she accepted and joined in (date).
Joan's project in my laboratory involved (describe in great detail, every once in a while adding one of the following sentences)

Her drive and determination were critical factors in the successful completion of a complex project that had proved too difficult for others in the field.

Her expert technical ability and keen intellect provided an ideal combination that allowed her to solve an important scientific problem.

These studies (have been / will be) published in (a high impact journal/ a journal that is widely read by those in the field).

---

[11] Address the letter to the person indicated in the job description. This is usually the search committee chair or the department chair. If a secretary or administrator, it is OK to use the generic "Dear Search Committee Members" but you should take the time to insert the proper address in the usual business letter format.

On a personal side, Joan is a delight to have in the laboratory. She regularly participates in helpful discussions and contributes to the projects of others in the group. Joan is much sought out for her knowledge of the field and gives expert advice. If we could offer her a suitable position at our university, we would be delighted to keep her. She is clearly one of the top 5% of all trainees of my (laboratory / department / university). You will be very fortunate to recruit her to your (department / university / institution) and are certain to be as impressed with her as we are. I recommend Joan with the highest possible level of enthusiasm.

Sincerely,
(your name and phone number[12])

---

[12] Some recruitment committees always call the references of the top candidates, so include your phone number in your letter.

The Importance of Collaborations

The public's image of a scientist is that of a half-crazed maniac working alone in a laboratory filled with flasks and beakers of brightly colored liquids bubbling over Bunsen burners. In reality, modern science is a team effort with large numbers of half-crazed scientists hunched over tables of boiling flasks and beakers. Scan through the pages of journals from 100 years ago and you will see many papers written by a single author, and most with three authors or less. Today, research papers with a single author are almost unheard of and those with only two authors are extremely rare, especially in the high-impact journals. To succeed in science today you need to be able to work in groups and contribute to a number of projects.

Setting up collaborations can be difficult—it isn't always easy to get others to help on your project. But usually the hardest part is at the end, once you begin to write up the project, and have to decide who gets listed as an author. Take the case of the traditional children's tale, The Little Red Hen, as modified for scientists.

The Little Red Hen

One day in the barnyard while the animals were talking about a new scientific result, the little red hen asked a question. None of the animals knew the answer. That gave the little red hen an idea. It was not a very large or well-developed idea, but none the less, it was an idea. The little red hen was worried that it was not a new idea so she asked all the other animals in the barnyard to help her dig through the literature to see if the idea had already been tested. One by one the animals responded. "Not I," said the turkey. "Not I," said the cow. "Not I," said the goose. "Not I," said the horse. And so on, and so on, until all the animals said no.

Well, thought the little red hen, I'll just have to dig through the literature myself. And so she did. Several weeks later she concluded that the idea was new—it did not appear in the literature. Now to develop the idea and figure a way to test it, she thought. Once again, the little red hen went to all the animals in the barnyard and said, "Who will help me grow this idea into a testable hypothesis?" And once again, the animals replied. "Not I," said the turkey. "Not I," said the cow. "Not I," said the goose. "Not I," said the horse. And so on, and so on, until all the animals said no.

Not a problem, thought the little red hen, I will simply do this myself. And so she did. She brainstormed night and day, often waking up in the middle of the night with a brilliant theory. Within a month she had grown the germ of an idea into a full hypothesis and sketched out a few experiments to test it. Full of excitement, she went back to the animals in the barnyard and said, "Who will help me test this idea?" And once again the animals replied. "Not I," said the turkey. "Not I," said the cow. "Not I," said the goose. "Not I," said the horse. And so on, and so on, until all the animals said no.

I guess I'll have to do the experiments myself, thought the little red hen. She worked very hard for nearly a year doing tasks that were nearly impossible for a chicken, due to the lack of fingers and an opposable thumb, but somehow she managed to pull it off and obtain the most beautiful data. Amazed at the high quality data, she went to the animals and said, "Who will help me analyze this data?" And once again the animals replied. "Not I," said the turkey. "Not I," said the cow. "Not I," said the goose. "Not I," said the horse. And so on, and so on, until all the animals said no.

At this point, the little red hen had suspected that this would be the response, but given how beautiful the data was, she had hoped that she could get help. Still, she went ahead and analyzed it herself, performing the necessary transformations, deconvolutions, and statistical analyses to test the significance of the findings. Finally she was done! And best of all, her hypothesis was correct! Now she could write a scientific paper and let the entire community know about her new theory. Once again, she went to the animals and asked, "Who will help me write this research article about my new discovery?" And once again the animals replied. "Not I," said the turkey. "Not I," said the cow. "Not I," said the goose. "Not I," said the horse. And so on, and so on, until all the animals said no.

The little red hen didn't bat an eye and went ahead and wrote the entire paper. And when it was ready to be sent out to the journal, she went to all the animals in the barnyard and said, "Who would like to be a co-author on the manuscript?" She expected to get the same response she had for all her other requests, and was completely shocked by the answers. "I will," said the turkey, "for I guided your research." "I will," said the cow, "for I was part of the discussion that led to the idea in the first place." "I will," said the goose, "for it was my previous work that you built your hypothesis on." "I will," said the horse, "for without me the barnyard would be dysfunctional." And so on, and so on, until all the animals said yes.

When the farmer heard about the research and the manuscript, he also insisted to be listed as a co-author because the barnyard wouldn't exist without him. And the farmer's wife also asked to be a co-author for the simple fact that the little red hen had not been cooked and eaten long ago despite a rather poor output of eggs. The little red hen listened to them all and then said with a laugh, "You didn't help with the development of the idea, the experiments, the data analyses, or even the writing of the paper so I'm not going to include you as co-authors. I will publish all by myself." The farmer's wife eventually found out and was so upset she served the little red hen for Sunday dinner.

Moral: (chose one)
- The ethically correct thing to do (i.e. not adding guest authors) is not always the most popular choice.

- If you are a hen, don't upset knife-wielding humans.

- If you are a farmer, take a more active role in your barnyard and help super-intelligent hens develop their ideas. Or better yet, make tons of money showing off your amazing barnyard of animals that talk, do research, and write papers.

Chapter 6
Moving Up

Soon after you've started your new position, you need to look ahead to the next level and have a firm understanding of what is expected for promotion. The emphasis on research, teaching, and administrative duties can vary from place to place, and even within a place, depending on your position. For example, if you were hired to run a centralized research facility, then the administration of this center is likely to be an important part of your eventual promotion. Likewise, if you were hired to teach an essential course or run a program, then you will need to focus on these duties. However, most scientists in academic settings, and even those in industry or government positions need to excel at research in order to move to the next level. In academic institutions in the US, it is common to require a national reputation for promotion to associate professor, and an international reputation for promotion to full professor.

In order to have a clear understanding of what is expected of you in your new position, it is a good idea to get advice from others in your institution. Ideally, you will find 2-3 career mentors—people who are more senior to you and can be consulted for advice from time to time. This isn't like your graduate thesis advisor(s), who met with you frequently and probably suggested specific experiments. Instead, your career mentors will give you practical big-picture advice on the appropriate focus of your time, whether on research or other duties. Mentors should also be consulted on grant applications, and anything else that is very important. Some universities have a formal mentoring process where senior faculty members are assigned to junior members of the department, but most places have only informal programs, if any. Even if you're not required to seek out mentors and have yearly meetings, it is an excellent idea to get advice from senior people when you are in the early stages of your career.

When first starting out, you need to focus on one or two projects and then gradually add on more projects as your lab expands. Often, the initial experiments in your own lab are to finish up projects you started when you were a post-doc. This is usually a productive way to start your lab's research program and get publications from your new place. However, to establish your independence you need to develop projects that will not include your previous mentor as a co-author when the results are eventually published. Many promotion committees will hesitate if all of your publications include your previous post-doc advisor as a co-author. Promotion from assistant to associate professor indicates independence, and you need to show this in your publication record.

The classic children's tale "The Three Little Post-Docs" nicely illustrates the dilemma faced by many post-docs. In case you were deprived of this story as a child, here it is again.

The Three Little Post-Docs

Once upon a time there lived three little post-docs who all worked in the lab of Dr. Wolfe. One day Dr. Wolfe said to them, "You have all been here for many years, and it's time you ventured out into the real world to find independent positions." The three little post-docs didn't want to leave the warm and comfy lab, but after Dr. Wolfe told them how much more money they could earn on their own, they hastily said their good-byes and trotted off in search of independent positions.

The first little post-doc to find a position and begin setting up his lab was Jack Straw. He hurried and scurried and in no time at all he had constructed a theory that looked quite good to the casual observer. It was not long before Dr. Wolfe heard of Jack's progress and paid him a visit. After Jack showed Dr. Wolfe his model and a draft of a manuscript describing his recent test of the model, Dr. Wolfe was impressed and asked to be included as a co-author since, after all, it was a logical extension of the project Jack had been working on in Dr. Wolfe's lab. "No way," said Jack Straw, "I'm independent now." Upon hearing this, Dr. Wolfe's eyes glazed over and he began to stroke the hairs on his chinny chin chin. "If you don't include me as a co-author," Dr. Wolfe shouted, "I'll huff and I'll puff and I'll blow your theory down." When Jack still refused, Dr. Wolfe began to rant and rave and in no time at all had pointed out dozens of fundamental errors that completely destroyed Jack's model. His confidence destroyed, Jack went running from the room.

Meanwhile the second little post-doc, Susan Stick, had also found a position and begun to set up her lab. She wasn't as hasty as Jack Straw and she took longer to come up with her theory, but when she did, this theory looked quite solid. Her colleagues agreed that she had built a fine model and should write a paper describing it. Even Dr. Wolfe said it looked good when he stopped by on a visit. "In fact," said Dr. Wolfe, "you should include me as co-author on your manuscript since, after all, you did start some of the work in my lab." "No way," said Susan Stick, "I started the project after I left your lab, and I'm independent now." Once again, Dr. Wolfe's eyes got that glassy look, he stroked the hairs on his chinny chin chin, and he shouted, "Include me as co-author or I'll huff and I'll puff and I'll blow your theory down." Susan hesitated—she had heard of what happened to Jack Straw, but she figured that her model was much more solid and would withstand any assault and so she refused to include Dr. Wolfe as co-author. With that, Dr. Wolfe began to rant and to rave and to rant some more. At first nothing happened—Susan's theory seemed to hold. This just enraged Dr. Wolfe even more and he ranted and raved even harder than before. Eventually he hit upon a small crack in the theory. It wasn't long before Dr. Wolfe expanded this minor oversight into a major flaw that seemed to jeopardize the whole theory. Her theory damaged, Susan ran from the room.

The third little post-doc, Bill Brick, eventually found a position. He was much slower that Jack and Susan and took his time to think things through before doing anything. After several years, Bill finally had constructed a theory and had some solid evidence to support it. Just when Bill was completing a manuscript describing his new theory, Dr. Wolfe stopped by. Needless to say, Dr. Wolfe was very impressed with Bill's theory, and he very much wanted to be part of Bill's paper. "No way," said Bill, "this was started well after I left your lab." "Yes, but the ideas were developed in my lab," said Dr. Wolfe.

After 30 minutes of bickering, Dr. Wolfe's eyes glazed over, he stroked the hairs on his chinny chin chin, and he gave the ultimatum, "Either include me as a co-author or I'll huff and I'll puff and I'll blow your theory down." Bill was well aware of what happened to Jack and Susan, but he was sure that his theory was watertight, so he politely refused Dr. Wolfe's offer of co-authorship. Dr. Wolfe began to rant and to rave and to rant and to rave, but to no avail. The theory could not be shot down. Dr. Wolfe intensified his attacks, trying every possible angle to find a flaw, even the tiniest one. He hopped up and down while thinking so hard that steam could be seen rising from his bald head. Suddenly, without warning, he stopped in mid-sentence and collapsed on the floor. He died before the ambulance arrived. The cause of death was found to be a ruptured blood vessel in his cerebral cortex.

Epilog:

Bill Brick gained the respect of his peers for standing up to the legendary Dr. Wolfe, but his slow and careful pace didn't produce enough manuscripts to sufficiently impress the reviewers of his grant applications. Without steady grant support, Bill was unable to get tenure. He struggled on for many years in a low paying soft money position, being forced to pursue minor projects with immediate results rather than use his creative talents to study matters of importance.

Susan Stick had a larger output of papers than Bill Brick, and most of her papers were high quality studies of importance to the field. She did quite well as an independent scientist, eventually rising to become chair of her department.

Jack Straw had a rapid rise to fame due to his very large number of high impact papers. However, things collapsed when many of his finding could not be replicated, and eventually he lost the respect of his peers.

Dr. Wolfe remained dead, although his legend lived on. Over time, the stories of Dr. Wolfe achieved a "Paul Bunyan" level of myth. Late at night when the moon was full and the wind was blowing hard, students and post-docs could hear his voice ranting away in the distance "I'll huff and I'll puff..."

Moral: (chose one)

Don't cry over spilt theories.

Slow and steady may win some races, but lots of solid publications are needed to get a grant.

Old professors never die, they just rave away.

The Phases of a Scientific Career

Age	Energy Level	General Outlook	Feeling that a Big Breakthrough is Just Around the Corner
Young	Enthusiastic	Idealistic	Hopefully optimistic
Old enough to know better	Realistic	Jaded, cynical	Optimistically hopeful
Wasn't born yesterday	Tired of the same old rat race	Skeptical	Hopeful but not optimistic
Old fogey	Enthusiastic, at least in spirit	Idealistic	Could you repeat the question?

Chapter 7
Making Groundbreaking Discoveries

All scientists recognize that there is an element of luck in the discovery process, especially the huge breakthroughs that cause paradigm shifts. As scientists, we are usually looking for things that are novel. We have ideas of new things, not yet discovered, which we predict to exist. But they may not—our hypothesis could be wrong. Or, they may exist but our method of searching is not correct. We can also be looking in the wrong place. To know where to look, and with what tools, takes some wild guesses. And to succeed you have to be lucky.

Luck is obviously not something you can simply purchase and add to your experiment. But successful scientists often do things to increase their chance of getting lucky (although few would describe it this way). These are not superstitious things, but practical things that simply improve the odds of discovering something important. These things are listed below, and are discussed in more detail in this chapter.

Ways to Get Lucky
 1. Be flexible in your approach

 2. Ask broad questions

 3. Work on things that are likely to succeed and lead to important discoveries

 4. Expand your search

 5. Consider what you've already found

 6. Avoid bias that limits your thinking

1. Be flexible in your approach

The discovery process is an exploration of the unknown. Scientists are just like explorers of past centuries who mapped out unchartered territories. A well known example is the first US expedition to the Pacific Coast by Meriwether Lewis and William Clark, along with dozen of soldiers and many local Indians who guided them (such as Sacajawea). The general instructions to Lewis and Clark were to head west. But rather than go west 100% of the time, they followed rivers and valleys on their route, at times going north, south, or even east! It didn't make sense to take a direct route going west and only west, over mountains and cliffs, especially considering the tons of supplies they had to carry. Their fastest route was to focus on the major goal— head west—but not if there was something in the way. They went *around* the mountains!

Likewise, in your experiments be aware of the overall goal and don't hesitate to take a detour to go around obstacles. I took this approach when doing experiments in graduate school, trying to discover the enzyme that produced brain neuropeptides such as enkephalin (the body's endogenous opiate). My initial studies in the brain were complicated by too many enzymes, so I tried a simpler system—the adrenal medulla (on the advice from a post-doc in the lab). There were fewer enzymes in the adrenal medulla than in brain, and I was able to detect a unique enzyme with the right activity to be the enkephalin-producing enzyme. But to purify this enzyme from the adrenal would have been difficult because levels were low. After testing several tissues, I found 100 times more enzyme in the pituitary, and had it pure in no time at all. Of course, I then had to show that this was the same enzyme as the one in brain and adrenal, which it was. But if I had stuck with the original plan of studying the brain enzyme and had not taken detours, first to the adrenal medulla and then to the pituitary, I might still be a graduate student!

As obvious as it may seem to go around obstacles, it is not an approach that all scientists follow. Some people insist on a single path, without deviation. On several occasions I have met junior scientists working in laboratories focused on purifying new substances from the brain. In two cases, the progress was particularly slow. I told the students my own story—the one described above—and they rolled their eyes; they had already thought that it would be a good idea to look in other tissues. However, their advisors prevented any deviation from the brain. In both cases, the students eventually went ahead, after years of frustration, and looked elsewhere. In one case, much higher levels were found in another tissue and the factor was purified a couple of weeks later. In the other case, there was absolutely no difference in levels of the activity in any of the tissues examined. But that was a problem. The expectation was that levels should vary among tissues—biological substances are rarely present at exactly the same levels everywhere. The student then performed a control containing only the extraction solvent, no tissue, and found just as much activity as the samples with tissue extract. It turned out the bioactive substance was an impurity in the methanol used to extract the tissue. Once they stopped adding the tissue to the solvent, using methanol alone as a starting point, it was relatively easy to purify and identify the compound. (And even though it was not endogenous, it still was an important finding as a new bioactive compound and the result was eventually published; this illustrates point #5, below.)

The bottom line is that if things aren't working one way, rather than just repeating over and over, consider taking a detour around the problem. Keep in mind the overall goal and alternative routes to get to this goal.

2. Ask broad questions

Another way to be efficient and increase the chance of success is to ask broad questions. Science is a lot like the game Twenty Questions, which was popular in the 1940's and 1950's as a radio and television show, and was played in my family during long car trips. In this game, someone thinks of an object and the others have to guess it by asking questions, no more than twenty, and limited to those where the only answer can be yes or no. To succeed, you can't start by asking specifics—is it *that* dog over there, or is it any dog at all. Both are much too specific. You need to start with broad questions; is it an animal? If the answer is no, then that rules out the specific dog you had in mind, as well as billions of other possibilities! After the broad questions, you then move towards more specific questions to narrow down the choices.

You should use a similar approach in science. If, for example, you search the genome and find a gene that has homology to a gene encoding a well-studied enzyme, and you want to figure out the function of the protein produced from the new gene, you can guess that its function is just like the well-studied homolog and design an experiment to specifically test this hypothesis. Or, you can start with more basic questions—where is it expressed and what are its enzymatic properties? If your hypothesis is correct, you will find that the new protein is expressed in the right place, and has the correct activity to fit your hypothesis. But if your hypothesis is wrong, you will be much closer to finding the new function than if you had only tried to confirm that your original hypothesis was correct. By pursuing broader questions first, you are still testing your hypothesis, but with a more general set of experiments that allow for more directions if the hypothesis is wrong.

3. Work on things that are likely to succeed and lead to important discoveries

It seems self-evident that we would choose to pursue projects that have a high potential impact and which are certain to succeed and easy to accomplish. Unfortunately there are few of these projects out there—if important stuff really were so simple to do, others would have done it. Or maybe not—it depends if anyone else thought of it. If you have developed a unique technique or reagent, it may be possible to do some simple things that will have a high impact. But in general, we are usually trying to find a balance between three main issues:

Easiness to test

Significance of potential results

Probability of success

Let's define these as E (easiness), S (significance), and P (probability). The priority of a project would be estimated by the equation:

$$Priority = E \times S \times P$$

Nobody calculates a precise formula for deciding which projects to pursue, as it would be impossible to predict the significance with accuracy, and we always seem to think things will be easier than they turn out to be. Still, considering these three factors is useful. If something ranks highly in all three of these categories, it is definitely worth pursuing.

If you can't find any projects that score highly in all three categories, do you go for the easy projects that will likely succeed but not have much significance? Or do you go for the easy stuff that will likely be groundbreaking, even though the probability of success is very low? Or try things that are likely to succeed and yield important results, but which are technically very difficult to accomplish? There is no right answer, although there is a wrong answer: working on difficult projects that probably won't work, and even if they do succeed, nobody will care about. Avoid these!

An optimal strategy is to have a balanced approach, working on some projects that are likely to succeed and lead to publications, even if not high impact, and also working on some riskier and/or more difficult projects that are likely to be important contributions to the field.

4. Expand your search

Making a big discovery is like winning the lottery. But how can you possibly increase your odds of winning the lottery? Simple—buy many tickets! If you buy 10 tickets instead of one, you will have a 10-fold greater chance of winning. In science, the analogy is to work hard and try as many things as possible. If it is only a little more work to test ten things at a time, rather than only one, do the larger study—you will have a greater chance of success. The recent explosion of the "omics" fields (genomics, proteomics, and many more) has followed this strategy. Rather than look at one thing, look at 1,000 things at once. Or more! High-throughput screening is another way to try many things, and combinatorial libraries are common in industry. Some large screens combine a number of compounds in each sample, with the idea that any positive sample will then be retested with each of the individual components in order to identify the active substance. Working faster, with assays that take less time, is another way to expand your search and screen more samples. Work hard, work fast, and work smart, as in the following song.

Lab Work (to the tune of "Rawhide")

Rollin' rollin' rollin', cell lines are a-growin'
Keep the data flowin', lab work!
We've got a data goldmine, a high-throughput pipeline,
In one day we can do what once took years.
So much calculating, the answer is awaiting,
Then we can go out and drink some beers.

{Chorus} Round it up, round it down,
        Round it down, round it up,
        Round it up, round it down, who cares!
        Write it up, write it down,
        Write it down, write it up,
        And publish, publish, publish, somewhere.

Data it's a movin', project it's a groovin',
Though some are disapprovin', lab work!
Low wages are we earning, and who knows what we're
        learning,
There's no hypothesis for us to test.
But even if it's rubbish, just be the first to publish,
And when we all get tenure we can rest.

{Chorus} Write it up, write it down,
        Write it down, write it up,
        Write it up, write it down, who cares!
        Send it in, send it out,
        Send it out, send it in,
        Just publish, publish, publish, somewhere.

5. Consider what you've already found

Most scientists are focused on research. But what does this word really mean? The prefix *re* means to do again, and *search* means to look for, not to actually find, but to look for. So the term *research* really means to look for something again. This isn't efficient. Think of how successful you would be if you didn't do research, but instead just searched once, found some things, and then spent your time characterizing them and writing papers!

The classic way of doing science has been to develop a hypothesis based on previous observations. Then, experiments are designed to specifically test this hypothesis. If the results prove the hypothesis correct, you publish (or at least, you write it up for publication and try to convince the journal editors/reviewers that you have a new and exciting story worthy of publication). But, if your experiments disprove the hypothesis, then usually it's back to the drawing board—it can be difficult to publish negative data even though this is important because it will guide other people from going down the same path.

While the classic scientific method is very useful, an exclusive focus on a single hypothesis can prevent one from seeing the true significance of the experimental results. For example, when Christopher Columbus first arrived in the "new" world (which wasn't new to the people living there) he thought he had landed on some islands off the coast of India and found evidence to support this theory—plants had fragrant odors that resembled spice plants, the natives appeared to resemble southeast Asians, and the numerous islands he encountered seemed to fit his maps of Asian islands. Of course he was nowhere near India—he was on some islands off the coast of North America, where the natives lived in a primitive stone-age society much different from the sophisticated culture of India in the late 1400's. A biographer of Columbus wrote, "It is curious to observe how ingeniously the imagination of Columbus deceived him at every step, and how he wove everything into an uniform web of false conclusions."[13] Columbus didn't see the problems with his hypothesis, and went back to Europe convinced he had found a new route to India. He completely failed to see the larger significance of the discovery—a new land unknown to Europeans.

There are countless stories of modern scientists making a similar error and not realizing the significance of their discoveries. While it is appropriate to do experiments to test a hypothesis, keep an open mind when interpreting the results and you just might discover something even more important than what you set out to find—something you weren't looking for because you had no idea it even existed. These can be the most important discoveries of all.

Be alert for the true significance of your experimental results

---

[13] The Life and Voyages of Christopher Columbus, by Washington Irving (1828), pg 128.

6. Avoid bias that limits your thinking

Most scientists think that they are not biased. But this is far from the truth. All humans have biases. We see something happen and then generalize to all things. It is part of human nature. Because you are probably skeptical that you have any bias, let me pose a riddle, slightly modified from the original version I heard several years ago.

Riddle: A mother and her daughter are seriously injured in a car crash. They are rushed to the hospital. When the injured child is wheeled into the emergency room, the nurse on duty sees the child on the stretcher and cries out, "Oh no, this girl is my daughter!" How is this possible?

Most people immediately come up with complicated explanations—the girl has two mothers for a variety of reasons: birth mother, egg donor, step-mother, adopted mother, mother-in-law, God-mother. When these choices are eliminated—the child has only one mother who was injured in the car accident and is not the nurse on duty—most people will then think it is a case of mistaken identity.

Why does the simplest answer evade us? Because of bias! We have seen nurses in real life and in movies or on TV, and they are almost always women. So we formulate a rule in our minds that nurses are females. This is obviously not true; male nurses certainly exist. But we don't think of this in our attempts to answer the riddle. Children have two parents. If the nurse is not the girl's mother, then the simplest explanation is that the nurse is the girl's father. It's a trivial explanation, but it eludes most of us because of our bias.

In science, bias is common. There are countless examples where dogma got in the way of the interpretation of the results, and only when the dogma was questioned could the results make sense. In some cases, I have gone to the primary literature to check the original report behind the dogma and found that only a small number of observations led to a grandiose generalization.

How does this fit into a chapter on getting lucky? Making a discovery isn't just about getting data—it is the interpretation of the data. In many cases, the person who made the groundbreaking discovery saw the same data as other scientists. The difference is that the person who puts it all together is the one who has made the discovery.

A great example of this is the discovery of natural selection by Charles Darwin. Many biologists before Darwin had noted that the fossil record from ancient times did not contain most of the present-day animals, but instead showed related species which were often much larger than the present-day forms. Some scientists tried to explain this by postulating that the bones expanded during the fossilization process. An alternative interpretation is that the animals which lived in ancient times and produced the fossil record are no longer around, and new species have taken their place. This seems rather obvious now (at least to scientists—there are many people in the US who still don't believe in evolution). But at the time, this idea was not obvious at all.

The most difficult bias for Darwin to overcome was the widely held belief that humans, as well as all animals and plants, had been created in their present day form. Granted, the church had clearly been wrong before on the "earth as the center of the universe" idea, but the creation of the world was not something to treat lightly. Darwin was well aware of this, having spent some time studying in Christ's College, Cambridge, to become a parson. But the transmutation of species, as evolution was called at the time, directly contradicted the church's views of creation and so Darwin hesitated. He almost hesitated too long, and was nearly scooped by a young biologist named Alfred Wallace who independently came up with the idea of natural selection while recovering from malaria during a specimen-collecting expedition in the East Indies. Wallace mailed his manuscript to Darwin, who rushed out his own manuscript and submitted them together to the same journal.[14]

---

[14] This is true. Darwin's opus "On the Origin of Species" gives extensive credit to Alfred Wallace for independently discovering the same concept.

The Fanny Letters of Charles Darwin

Did Darwin really come up with the idea of natural selection on his own? Charles Darwin had a girlfriend, Fanny Owen, in England when he sat sail for South America. Charles and Fanny corresponded by letter during the early part of his journey, at least until she found another suitor and got married soon after Darwin left for his "2 year" journey around the world (which, as typical for a scientist, ended up taking a little longer than planned—4 years, 9 months, and 5 days—it wasn't just his dinner that got cold, but the entire relationship!). Here are the infamous Fanny letters, made public for the first time.[15]

March 1, 1832
Bahia, Brazil

Dearest Fanny,

Thank you so much for your letter, which I received when we arrived in Brazil. After months onboard the Beagle, I managed to stroll the forests of the new world in earnest. I wish I could more fully explain my research to you, but I fear that I could not put it into words that would make sense to a person untrained in science. In your letter, you mentioned that our relationship must "evolve" and you also used the term "evolution." I am unfamiliar with these words and assume it's a new concept that has come into usage since

---

[15] Not true—this was previously published: Fricker, L.D., The Darwin letters, Einstein Quarterly Journal of Biology and Medicine, 19, 47, 2002. But Darwin really did have a girlfriend named Fanny who ditched him after he left for his "short" cruise—at least that part is true, even if the letters have been made up!

my departure. What is meant by evolve and evolution? From the context of your use of the word, I would guess that evolution means gradual change or adaptation, perhaps into something better. I think it a catchy word although I doubt it will ever become commonplace.

I have given considerable thought to the question you raised in your letter regarding the origin of species. I had not previously considered this question, assuming that the Creator had simply been responsible for everything. But now that you've raised this issue, I've been looking at my data with a different angle. I will try to explain this more fully, as best I can to a person untrained in modern science, once I have completed my studies.

It is time for the supply boat to return to England and so I must end this letter. I eagerly await your next correspondence. Even though your letters are a diversion from my scientific research, I deeply enjoy them. Until your next letter arrives I will give considerable thought to your questions and this concept of evolution that you raise, and will hopefully be able to reply more fully to you in my next letter. Although other scientists scoff at the idea of trying to explain scientific ideas to a female, and some even remain bachelors so that they can focus completely on their work, I find your questions to be stimulating. Eventually, I hope to explain my research to the public and so it is important for me to try to explain it to you in terms you can understand. Perhaps I can even incorporate some new terms, like this concept of evolution that you speak of.

Yours affectionately,
Chuckie Darwin

July 23, 1832
Rio Plata, Brazil

Dearest Fanny,

Your letters are a constant stream of happiness without which I could not possibly endure the months of hardship on the H.M.S. Beagle. I'm nearly done with my collection of specimens from the coast of Brazil, and will soon travel to Patagonia. During my excursions, I've been mystified by the large diversity of species on this planet. It is puzzling as to why the Creator would generate so much variety of wildlife in each place we visit and it would seem to have been much more efficient to create the same species throughout the world.

In your last letter, you raised several issues that are new to me. I had not realized it before but now that you mention it I see that we have both changed during our separation. It has not been any one major event but little by little over time we have grown more different, or "adapted to our new environments" as you say. Your reference to our relationship, and relationships in general, as "surviving if they are fit" is most astute. This concept of "survival of the fittest" is most certainly true of relationships. It's too bad that such simple concepts do not apply to my studies for it would appear that the natural world is so much more complicated than relationships between close friends.

Another term you used in your last letter, "natural selection," is not at all clear to me. Do you mean to say that relationships are selected by natural events? I most certainly hope that our relationship will endure the two years of my absence, and that no further selection will take place. I count the days until we are reunited, and I can put these silly notions like natural selection out of your mind.

I have enclosed another sketch of myself done by a crew member who is artistically inclined. In your letter, you commented that the last sketches I sent remind you of a monkey. I did not notice before, but now that you mention it

I can certainly see the resemblance that is even more striking in the present sketch. It's been months since I shaved or had a decent haircut. Life on the boat does not place much emphasis on the grooming that is so customary back home. It is most interesting that in the absence of the influences of society, man adopts the appearance of a monkey. I feel that this is somehow relevant to my own research on the characterization of species, but can't quite put my finger on the connection.

Once we have visited Tierra del Fuego, it will be a simple matter to complete our circumnavigation and I shall certainly rejoin you before another year has passed.

Yours affectionately,
Chuckie Darwin

October 12, 1832
Montevideo, Uruguay

Dear Mrs. Biddulp:

It has come as quite a shock to learn that you have married, and it will take me some time to consider you as anything other than my dearest Fanny. I see now in your previous letters that you hinted at the possibility of ending our relationship and created terms like "natural selection" and "survival of the fittest" to rationalize your actions. If it were not for my absence I doubt this would have been the course of events, and I regret that I embarked on this adventure.

As this is the last letter I will write you, given the circumstances, let me wish you well in your new life. I will always cherish the letters you have written. I feel that some of the points you have raised may be of broader significance than just a simple relationship, although I do not understand it yet.

As you surely realize, there is little opportunity for me to purchase an appropriate wedding gift in my present environment. Please accept my humble gift that is enclosed. These are quite rare specimens of fossilized bird droppings that I collected with great effort and I hope you will appreciate the significance of my meager gift.

Sincerely,
Charles Darwin

Summary

To re-cap, these are the key points of this chapter (and the simple way of thinking about each):

1. Be flexible in your approach (go around the mountains)

2. Ask broad questions (play 20 questions)

3. Work on things that are likely to succeed and lead to important discoveries (ESP)

4. Expand your search (buy more lottery tickets)

5. Consider what you've already found (avoid "re-search")

6. Avoid bias that limits your thinking (think outside the box)

Chapter 8
Scientific Writing—the Key to Success

Writing is an important part of most scientific careers. Without publications, it is difficult to get funding and virtually impossible to get promoted unless your job involves mainly teaching and/or administrative duties. Publications are the life-blood of science. Until a result is published, it is not considered to have been discovered; the date of the publication is often said to be the date of the discovery. Most importantly, credit for a discovery usually goes to the first group to publish, not the first group to actually make the discovery. For example, while Alexander Fleming is most commonly credited with discovering penicillin, Ernest Duchesne discovered the antibiotic properties of penicillin extracts several decades before Fleming but didn't publish his findings (except in his thesis), and then dropped out of research to be a physician. On the other hand, it wasn't entirely Duchesne's fault—he did try to convince the medical establishment but they were not interested. But this is a common problem with paradigm shifting discoveries, and many important discoveries required years of work outside the establishment until skeptics were convinced and the idea became widely accepted.

Once you have published several key research papers in an area, consider writing a review on the subject—this can look great on your CV and will help you get established as a leader in the field. For example, consider the co-discovery of natural selection by Charles Darwin and Alfred Wallace; both published their initial report at the same time, and in the same journal. So why do most people only know the name of Darwin, not Wallace? Both scientists stayed in the field and went on to write many more papers on the subject. Darwin got the credit for the discovery, in part because he really had the idea long before he published and had talked to friends about it. But the main reason Darwin went down in history was because he wrote the book on the subject. Even though Wallace was cited as a co-discoverer of the theory of natural selection in Darwin's book, because Darwin was the sole author he became known to the public.

While writing entire books in a scientific discipline is a bit rare, you should consider writing reviews from time to time. Although many reviews are solicited by the publisher, there are some journals that allow for people to submit ideas. Promotion committees on which I have participated like to see reviews written by the candidate—this shows that the person is a leader in his/her field.

The Art of Writing Scientific Papers

Scientific writing is fairly easy to learn but hard to master. There are general rules to writing a scientific report. It's not like creative writing where the limit is your imagination; science writing involves describing a set of observations and your interpretations as to what they mean. Once you learn the basic rules, which you should do as a student and post-doc, it should be relatively easy to write a draft of a manuscript. It may take you considerable amounts of time, but as you get more experience the entire process will become more efficient. In addition to writing a draft in shorter time, you will learn how to write a better draft—one that will need less editing and have a stronger chance of getting accepted for publication. For this, it is important to consider the big picture. Literally!

All writing, both scientific and non-scientific, should have a clear point for the reader to focus on. One way of thinking of this is to look at art, especially Renaissance paintings. For example, Leonardo da Vinci was a master of perspective (among other things). His paintings had a clear focal point, whether simple (Mona Lisa) or complex (The Last Supper). In the latter example, there are many details in the painting that all fit together into the overall theme, but the viewer is drawn to the main element (Jesus) in the center of the picture. Likewise, a well-written scientific paper can include a number of sub-themes that relate to the main idea, as long as these side points complement the central theme without diverting the reader's attention from the main focal point of the paper.

Leonardo's "Mona Lisa" is a good model for a scientific paper with a clear focus on a single object.

Leonardo's "The Last Supper" is a good model for a scientific paper with many details and side themes, but with a clear focal point. Note the way the viewer's attention is drawn to the central object (Jesus) by a variety of techniques such as the use of light and lines of perspective (focal point of the room, finger pointing and gaze of the Apostles).

In a scientific paper, you want your focal point to be the discovery you have made. Don't draw too much attention to the unknown—it is OK to mention some areas for further experiments but this should not be the main theme/focal point.

While many pieces of art are excellent examples of how to focus a scientific paper, there are also examples in the art world of what *not* to do. For example, the paintings of Jackson Pollock are beautiful and captivating, but generally lack a clear focal point; this is not appropriate for scientific writing. The idea is not to simply dump information in a paper but to present a clear picture of what you have found and the importance of this discovery.

An imitation of a Jackson Pollock-style painting (by Arun Fricker when in preschool). This is possibly great art, but certainly not a great example for scientific writing; there are too many distractions without a clear focal point. The eye is drawn to many points within the painting.

Another example of what to avoid in writing scientific papers is the modern art painting "Black Square" by Kazimir Malevich, which lacks all detail, focal points, or anything other than black paint! Avoid this in your writing, except maybe for committee reports.

While there are some parallels between art and science, there are obviously differences. Modern art is not something the average person can understand—it is strange and abstract. On the other hand, modern science is … well… OK, so the average person also finds modern science strange and abstract. But it isn't. At least other scientists understand it. Some of the time!

Getting Your Work Published

It is not enough to write a manuscript describing your results—you also have to get it published. There are three things you can do to help improve the chances that your manuscript will be published.

1) Describe the importance of your discoveries—don't assume the reviewers will see the significance.

2) Keep it focused. When you write, tie everything together into a single story with a beginning, middle, and end. The beginning is the question you set out to answer, the middle is your results, and the end is the answer to the question. If you can't tie everything together into a story, then reconsider the data you've chosen to present in the paper and either remove what doesn't fit or go back to lab to do the missing experiments that will connect everything.

3) Don't give up if a manuscript is rejected. You should consider the reviewers' opinions—maybe they have some valid points that you can address. But keep trying to publish.

A common mistake is to look at the lengthy list of issues raised by the reviewers and decide that the only option is to send it to another journal. If you can respond to the comments, preferably by making changes in the manuscript, then you can resubmit your manuscript to the same journal. You don't usually have to do everything the reviewers ask for, but you should be able to explain why in your rebuttal letter. Many times reviewers ask for a considerable amount of additional work. If these are important controls, you probably need to include them. But if they are studies that will take many years of effort and result in a paper with 20 figures and tables, you can argue that these are excellent suggestions for further research which are well beyond the scope of the present manuscript.

If your work is rejected by a journal and the editors will not consider a revised version (or if they reject the revised version) then you can either scrap the whole idea, going back to the lab to do more experiments, or you can try another journal. If you decide on the latter option, it is rarely a good idea to completely ignore the previous reviewers' comments and send your manuscript to a second journal without making any changes—there is a good chance that the new reviewers will see the same problems that the first group of reviewers saw. Therefore, even if you are not resubmitting to the same journal, at least try to improve the manuscript by carefully considering the reviewers' comments before sending it to another journal, or you might find the same criticism from the new reviewers.

Can't Publish
(to the tune of "Can't Touch This" by M.C. Hammer)

My work is going, so strong
Makes me think it won't be long,
Until they are honoring me
With a Nobel Prize, physiology.
But before I can win that prize
My work I must publicize,
Cuz there ain't no discovery
Unless the whole wide world can see.
So I write it up, send it to *Science*
Many weeks pass and only silence.
Finally I get an e-mail
Rather short, without much detail.
It says "Dear Sir, we do thank you
For the chance we had to review.
But we get so many more
Submissions than we have the space for.
You are someone that we don't know,
So we'll have to let this one go.
I know that you will think we're clubbish,
But with us, you just can't publish!"

Chorus sings part in parentheses:
(You can't publish) But it's a big breakthrough
(You can't publish) This is Nobel Prize material!
(You can't publish) You're gonna regret this when I'm
famous (You can't publish!)

Years have passed, I'm getting old,
The data's getting rather cold.
Hundreds of journals I have tried,
But each time I've been denied.
They say not enough interest.
Now I'm mad, I'm thoroughly pissed.
Those hack reviewers are killing me.
They trash my work so willingly.
Those idiots are not my peers,
They're anonymous volunteers.
Pseudo-scientist wannabees,
Can't see the forest for the trees.
In desperation I decide
That one last journal should be tried.
An on-line journal, not in print,
"We publish all" or so they hint.
Their hefty publication fee
Is 'pay to publish,' I agree.
But as it seems this is my fate
I send it in and then I wait.
Weeks go by, months go past,
Then the answer comes at last.
 "Please don't ever resubmit
Cuz your works a piece of **** (*beep*)
What we mean is it's pure rubbish,
It's so bad—you'll never publish!"

(You can't publish) You don't know what you're missing
(You can't publish) This is hot stuff
(You can't publish) But I won't get tenure without
publications! (You can't publish!)

Most scientists regarded the new streamlined peer-review process
as "quite an improvement."

Chapter 9
Getting Funding For Your Laboratory

Every scientist running a research laboratory recognizes the importance of funding. Without it, one cannot do much research, and there never seems to be enough to pursue all of your interesting ideas. Funding, or rather the lack of enough funding, is the most common complaint among scientists. It seems like a large amount of our effort is directed towards getting more funding. At first, we are ecstatic when we get our first grant funded. Then we realize one grant isn't enough, and try for a second, and maybe a third and fourth. It never ends.

There are many books that give solid advice on writing scientific grant applications, including one in the same style as the present book which mixes serious advice with humor.[16] To summarize, the three key points to writing a successful grant application are

-Seek advice from colleagues throughout the process, from the initial planning stages to the final draft.

- Start the process early—months or even one year before the planned submission date so that you have time to get advice from colleagues and obtain the appropriate preliminary data to support your ideas.

- Propose something that is exciting but not so completely novel that the reviewers will have a hard time accepting the concept, even with preliminary data to support your idea. Most Nobel Prize-winning discoveries were not initially funded by grants for that project.

---

[16] How to Write a REALLY Bad Grant Application (and Other Helpful Advice For Scientists), by Lloyd Fricker (2004).

The Big Picture of Writing Grant Applications

In the previous chapter on writing papers, the concept of focal point was discussed. While both papers and grant applications should have a clear focal point that draws the reader's attention, the focal point is not the same. As mentioned in the previous chapter, the focal point of a research paper should be like Leonardo's paintings "Mona Lisa" or "Last Supper," with the emphasis on the main finding of the paper. In contrast, when writing a grant application you should draw the reviewer's attention to something in the distance—what you will certainly discover if the reviewers give your application a fundable score. Otherwise, if the focus is on what you have already done (Mona Lisa), then the reviewers will question the need for further experiments. Instead, you need a cohesive picture that is well balanced and with a focal point slightly in the distance.

A great example of the ideal balance for a grant application is the photograph "Long Island 1982." In this photograph, the viewer is drawn to the distant buildings by the converging lines of the pier. This focal point is visible but somewhat mysterious. Similarly, a grant application needs to draw the reader in and focus on something that is important but not well defined.

"Long Island 1982" (photo by the author) has a focal point in the distance, which is an example of the ideal focal point of a grant application.

The Importance of Being Persistent

Even if you write the greatest grant application ever written, it may not get funded. There are too many variables in the process. The main variable is the reviewers. If they are outside your field, they may not appreciate the significance of your proposal, even though you mention it in your application. Most of us think that our own field is one of the best areas for research—that's why we chose that field. And, the techniques we use are the best way to study the problem—that's why we're using them. But often, reviewers in other areas can't fully appreciate the value of research outside of their own field. Conversely, if the reviewers are directly in your field and using similar techniques, they will appreciate the significance of the research but may expect to see technical details that you didn't mention (because of the page limit!) and then fault you for that.

The solution to the randomness of the process is to do two things. First, spend time working on the application. Get advice from colleagues who can see it from different perspectives. Try to write the best application you can, targeted to the level of expertise you expect will be present on the panel that will review your application.

Second, and equally important, don't give up trying to get funding. Be persistent! You should consider a variety of funding sources; federal agencies, private foundations, and companies are common sources of research funds. Most schools have an office of grant support that can provide information on potential funding sources. The bottom line— don't be too discouraged if your first attempt at a grant application isn't funded. Keep trying, as in the following story.

Dr. Goldi Lochs and the Three Reviewers

*D*r. Goldi Lochs was doing an experiment and she stumbled upon an unexpected result. "This is odd," she thought to herself, "this result is completely new." She decided to delve into the field and check it out, and so she spent the afternoon reading papers and thinking of how her new result could be explained. Based on the previous literature, Dr. Lochs found three models. The first was very complex, with lots of arrows and hypothetical connections. "Too convoluted," she thought. The second was very simplistic. "Maybe too simplistic," she thought, "for nature is always more complicated than this." The third model was just right—not too convoluted, not too simplistic. "This is perfect," she thought.

Then she needed to come up with a hypothesis to explain her data in relation to the new model. "Well," she thought to herself, "there are three hypotheses that are possible." The first was a general hypothesis that fit many different models. "Too broad," she thought. The second was highly focused on just one aspect of the model. "Too narrow," she thought. The third was just right—it covered all parts of her model without being so general that it applied to everything. "This is perfect," she thought.

"Now I need to come up with an experimental model system to test this hypothesis." She came up with three approaches. The first was very difficult to set up and required years of training. "Too complex," she thought. The second was very easy to set up but provided only a little information. "Too limited," she thought. The third was less complex than the first but was not too difficult to set up and provided more information than the second experimental system. "This is perfect," she thought, "and I should write a grant application on this." She thought of three Specific Aims. The first Aim was a bit boring but provided an important answer that needed to be addressed, and the experiments were virtually guaranteed to succeed. The second Aim was a bit more exciting, but also less likely to succeed. The third was quite risky but if it worked, it would cause a paradigm shift in the field. "These are perfect," she thought, "just the right balance."

Dr. Lochs wrote the grant application and submitted it to a federal funding agency. She began to daydream about how she would spend the money and do great science. Just then, she was roused out of her daydream by her computer reporting that she had received a new e-mail. It was an announcement that her grant review was available on-line. She hurriedly logged into the site and eagerly read the summary statement. The first reviewer was very positive. "This is very exciting," the reviewer wrote, "but it is a bit too ambitious." The second reviewer was also positive and thought it was exciting but asked, "Where are data showing it will work? This is too preliminary!" The third reviewer was also positive and thought the idea was very exciting. But unlike the other two reviewers, the third reviewer thought it was just right! However, the average of the scores from the three reviewers was not enough to give the grant application a fundable score.

Dr. Lochs was heartbroken, but she didn't give up. After reading the reviewers comments and talking it over with colleagues, Dr. Lochs revised the application, got more advice from colleagues, made more changes, and then resubmitted the application. And, this time it was considered perfect by all three reviewers.

Moral: If at first you don't succeed, try and try again, unless it is a grant application to the US National Institutes of Health, at which point you are currently[17] limited to only two attempts before you have to entirely rewrite the project, propose completely new experiments, move to another state, change your name, and undergo plastic surgery![18]

---

[17] This is the current system in 2011—hopefully it will not last much longer!

[18] Just kidding—there's currently no requirement for plastic surgery.

An alternative idea for obtaining research funding

Chapter 10
Managing Your Time Wisely

When first starting out as the head of your own lab group, it is easy to become overwhelmed with all of the things that need to be done. You have to learn how to be efficient. Some businesses hire consultants to improve the efficiency of their workers. A friend who worked as a business consultant on efficiency summarized her company's advice as follows. First, sort your assignments into three piles. The A pile needs to be done right away, the B pile needs to be done but can wait, and the C pile doesn't ever need to be done. Throw this C pile away, save the B pile, and begin work on the A pile. When you're done with the A pile, move to the B pile. In other words, go from deadline to deadline doing what is urgent. One popular advice book for scientists suggests prioritizing your time by considering importance in addition to urgency.[19] The advice is to focus on the tasks that are both important and urgent, and not just those that are urgent but of little importance.

	Important	Not Important
Urgent	Urgent and important—make these tasks your top priority. Spend enough time to do these tasks well.	Urgent but not important—ignore if you can. If you can't ignore, spend as little time as possible.
Not Urgent	Important but not urgent—these tasks should be your second priority.	Not important or urgent—don't waste time on these tasks.

The problem with considering urgency is that for scientists, most of the very important things do not have rigid deadlines. For example, it is essential that we do experiments and publish our discoveries but we rarely have a deadline for submission of the manuscript (unless it is an invited paper

---

[19] At the Helm: Leading Your Laboratory by Kathy Barker (2010)

that is part of a book or special issue which has a firm deadline). Therefore, the important tasks should take priority over those that may be urgent but which are not important for your career.

Another way of grouping tasks is to consider urgency as part of importance; things that are important include projects with deadlines (grant applications, giving lectures at scientific conferences) as well as those without deadlines (publications). The second factor to consider is whether you enjoy doing it; the "fun factor." Your priority will likely be things you think are fun to do, but of this group you should spend most of your time on the important tasks and try to avoid spending too much time on things that are fun but not so important. Your second priority should be important things that you don't enjoy as much. Use the fun-but-not-important activities as a reward to yourself when you've completed all of the important tasks.

	Important	Not Important
Fun	Important and fun to do—these will likely be your top priority because you enjoy them.	Fun but not important—these tasks should be your third priority.
Not Fun	Important but not fun—these tasks should be your second priority.	Neither important nor fun—don't waste time on these tasks.

Good Enough Can Be Good Enough!

When my son "graduated" from elementary school, there was a ceremony complete with a keynote speaker who gave a 10 minute lecture to the bored kids and their parents on the topic "good enough isn't!" He meant well. Later in life, don't be a slacker who tries to get by with the minimum effort required—always strive for excellence at all things. But either the speaker lives in a different world than the rest of us or he is not very efficient with his time. Even a non-scientist has many mundane chores in life—grocery shopping, for example. If you go to the store, buy everything on your list, and return home with food, you did a good enough job and that is sufficient. Likewise, when filling out paperwork for various things, doing an adequate job is certainly good enough.

If you live by the motto "good enough isn't" then you will spend far too much time on the little things and not have enough time for the truly important tasks. Also, if you do a great job at things you really don't enjoy, chances are that you will be asked to do more of those things, whereas if you do a bad job you won't be asked again. But, if important for your career and promotion (teaching, administrative duties) you don't want to do a bad job, so spend just enough effort to do a good job on important things you don't enjoy. Then, you will have more time to focus and do an excellent job on what is both important and fun. That way if you are asked to do more of these things, you won't mind and will actually enjoy it.

Manage your time wisely—don't pursue experiments that aren't working...

Committees

Unless you are very lucky, you will have to serve on one or more committees. This can look good for your eventual promotion—it shows your dedication to the department and institution. Be careful in choosing which committees you join (if you have a choice—many places dump the worst committee assignments on the most junior members of the department). Some committees are quite important and necessary for the function of the university, while others are completely useless. Even the committees that serve an important purpose for the institution can be a waste of your time, and you have to learn to be efficient in dealing with the committee workload so that you can focus most of your efforts on your science.

Some people think there are three types of committees: useless, pointless, and a complete waste of time. But in reality, the third group is a large waste of time, not a complete waste!

Committees where people just talk. Most of these are useless, although some can be informative to you or others. For example, serving on an advisory committee for a student in another lab will help you learn what is expected of students in your new place, and you can possibly give scientific advice that will help the student. In contrast, faculty meetings that involve griping about the problems in your department for 1-2 hours, with no talk of solutions, are completely useless. If you can't avoid these meetings, at least try to turn the conversation around to consider solutions rather than merely complaining about problems.

Committees where people talk and then vote, but the vote doesn't matter because the chair will do whatever he/she wanted to do in the first place. These are usually completely pointless. If you are forced to serve on this type of committee, minimize the amount of time you spend in nonproductive conversations and hope that the meeting will end early.

<u>Committees that vote, and the vote actually counts.</u> These often fill an important role. At many places, the selection of students to admit into the program is decided by a committee. Also, decisions involving the hiring of new faculty and promotion are usually made by committees where the voice of each member matters (although not always equally). For some scientific societies, the program is selected by a committee; serving on one of these is a great way to get established in the field, or if you are already established, to help guide the field. Committees that actually accomplish something can be useful to serve on, but as with the others, try to be as efficient as possible so that you can devote the maximum amount of time to your scientific research.

Committees are rarely efficient...

Reviewing Stuff

When you first get asked to review a manuscript or grant application, it feels like an honor; you are an "expert" whose opinion matters. It can also be a great learning experience—your own manuscripts and grant applications will be more likely to get great reviews if you think like a reviewer when writing and editing your work. But it can take considerable time to review things, especially if they are not in your field. Don't hesitate to say no to reviewing things. Keep in mind the time that it will take away from your other duties. The criterion that I use to decide whether to accept a reviewing assignment is whether it is something that I want to read and/or will learn from. If not, I say no.

Even if you are selective in what you agree to review, it can take considerable time to read the manuscript/grant application and write a review. Some reviewers point out every typographical error. This is not appropriate—most journals employ copy editors to do this. A scientific reviewer should comment on the big picture; does the study advance the field, are there critical experiments that are missing, and have the authors avoided over-interpreting their data.

In addition to writing comments for the authors, you need to recommend to the editor whether the journal should reject the manuscript, request revisions, or publish without changes. To help distinguish these categories, consider the following criteria.

That's a Reject (to the tune of "That's Amore")

When the data don't say what the authors convey,
That's a "reject."
When the hypothesis is so surely amiss,
That's a "reject."
If the significance is no better than chance,
It's a "reject."

If the claims are not true and there's nothing that's new,
I say "reject."

When the dataset size doesn't convince my eyes,
I say "revise."
When the wording's awry and they must clarify,
That's a "revise."
If a critical test wasn't done, it is best
That they revise.
When their citation list's not complete (mine were
    missed)
I say "revise."

When the findings are new and there's no more to do,
I say "publish."
When they cite every one of my publications,
I say "yes!"
If the study's complete and the data looks neat,
That's a "publish."
When the sun and the stars are both lined up with Mars,
I say "publish."

{key change, then final stanza}

When I haven't a clue what they're trying to prove,
It's a "reject."
When there's no replicate then I don't hesitate
To say "no."
If the findings are old then I will be quite bold
And say "reject."
If the author's a jerk and once trashed my best work,
Then it's "reject."
{ending} A reject? Yes, reject!

E-Mail and Other Time Wasting Activities

Paradoxically, the biggest time-saving and time-wasting activity is e-mail. In the olden days, soon after the demise of the dinosaur, we would spend hours composing letters, place those letters into envelopes, and then put them into the mail where they would be carried on horse-back to the addressees. With the advent of express mail, and then the fax machine, the letter would get to its destination quicker, but we still spent considerable amounts of time composing actual letters. E-mail has saved some of this time by allowing for shorter transmissions that take less time to compose and send. But, in the old days we would pick up the telephone when we needed to have a conversation with someone, and this would usually take about 5-10 minutes to resolve the problem. Now, we can easily spend hours typing multiple e-mails to replace that 10 minute phone call. Take, for example, the following exchange, pulled from the internet using highly sophisticated computer hacking techniques.

To: smith228@aol.com
From: jones731@hotmail.com
Subject: Your elegant paper

Dear Dr. Smith,
Great job—your latest paper is one of the most important papers of this past decade, if not the past century. It certainly will be a landmark. You have resolved a long standing mystery that has evaded the top scientists in the world, and your brilliant solution to the problem is so elegant in its simplicity.

Your paper is so nearly perfect that I almost hesitate to bring this up, but I couldn't help noticing a very minor oversight. May I be so bold as to mention that you somehow overlooked citing a paper of mine that could be considered to be slightly relevant to your own work? In fact, some may even go so far as to say that your work was a logical extension of my previous study. I'm sure this oversight was a simple typographical error, and I must compliment you again on an extraordinary job.

Best Regards, Dr. Jones

To: jones731@hotmail.com
From: smith228@aol.com
Subject: You flatter me

Dear Dr. Jones,
Thank you so much for your flattering e-mail. It is an honor to receive such warm praise from a well-established scientist. My own contributions to the literature are surely not as important as your numerous and insightful publications. Regarding our choice of citations for the bibliography, I can assure you that the selection was based purely on the relevance to our experimental results. Each potential citation was carefully considered, and no typographical errors were made. I welcome your honest comments and hope you will understand that we simply could not cite every previous paper published.

Sincerely, Dr. Smith

To: smith228@aol.com
From: jones731@hotmail.com
Subject: Are you serious?

Dear Dr. Smith,
You are absolutely right in that one can't cite every paper in the field. However my previous paper, which you blatantly failed to reference, is more than just a seminal paper. It appears that you directly copied several figures and tables

from my previous work and published them in your own manuscript. If this doesn't qualify for citation, what does?

Best Regards, Dr. Jones

To: jones731@hotmail.com
From: smith228@aol.com
Subject: Lucky break

Dear Dr. Jones,
Your pathetic attempt at science does not qualify you for citation in anything. Your study was so full of errors that they actually canceled each other out and by sheer luck you arrived at the right answer. We spent a great deal of effort trying to replicate your experiments. Only after we started fresh, completely ignoring your so-called findings, did we get our experiments to work. Your publication set us back many months. Our deliberate decision to exclude this poor excuse for science from our citation list was an act of charity, for if we had, we would have needed to point out the numerous flaws that pervade your entire paper.

Sincerely, Dr. Smith

To: smith228@aol.com
From: jones731@hotmail.com
Subject: Your reputation

Dear Dr. Smith,
How dare you accuse me of poor science. You are the one with a reputation for results that are politely described as 'non-reproducible.' I'm surprised that you have not yet been investigated for outright fraud. In fact, I even doubt whether

the tabloid newspapers would accept your outlandish claims that are not supported by a shred of solid evidence.

Regards, Dr. Jones

––––––––––––––––––––––––

To: jones731@hotmail.com
From: smith228@aol.com
Subject: My reputation?

Dear Dr. Jones,
Are you talking about me, or yourself? I've overheard people at meetings questioning how you have survived in science all these years. Some even question whether you frequently use hallucinogenic drugs.

Sincerely, Dr. Smith

––––––––––––––––––––––––

To: smith228@aol.com
From: jones731@hotmail.com
Subject: Logic-impaired

Dear Dr. Smith,
I feel that we should not discuss science any further. You are hopelessly confused and deranged. I suggest you consult with your psychiatrist about increasing the dose of your medication.

Regards, Dr. Jones

––––––––––––––––––––––––

To: jones731@hotmail.com
From: smith228@aol.com
Subject: Truce

Dear Dr. Jones,
I agree that we should not discuss science anymore. By the way, isn't it your turn to cook dinner tonight?

Your loving husband, Dr. Smith

———————————————————————

To: smith228@aol.com
From: jones731@hotmail.com
Subject: Dinner

Dear Dr. Smith,
I believe it's your turn. I cooked last night.

Your loving wife, Dr. Jones

Another time-waster: trying to figure out what to do when the instructions are not clear.

"I'm beginning to wonder if our Terms of Reference may be a little too broad..?"

Chapter 11
Attending Conferences—Having Fun
While Advancing Your Career

Attending scientific conferences is important at all stages of your career. Scientists need to attend meetings to learn about new developments in the field. Although we spend a considerable amount of time reading scientific papers from other groups, it's not the same as hearing a seminar—often scientists include unpublished observations in their talks (or at least, they should!). Although most presentations at meetings focus mainly on the results of experiments that worked, the information provided in the question/discussion period after the seminar often provides information on what hasn't yet been done, or what has been tried but did not work. On many occasions I have saved months or years of work, or helped others save time by avoiding experiments that have already been done and are about to be published, or which have been tried but did not succeed. Also, at meetings we can learn about things in different research areas that impact on our own work but which are published in journals we don't normally look through.

Another equally important reason to attend meetings is to meet colleagues in the field and establish a rapport. Junior scientists need to become known in the field, and need to meet senior people who are the current gate keepers—the people who review submitted manuscripts and grant applications. It is also important for junior scientists to meet others at the same level—these people will eventually become the peers of the peer-review process. Although friendship does not guarantee that a manuscript or grant application will be given a favorable review, it certainly helps to have friends review your work rather than having strangers do this (or even worse, enemies). Competition is inevitable, but it is often possible to turn competition into collaboration over a beer or two at a meeting. Or at least you can make it a friendly competition rather than a fight to the death.

The third important function of meetings is the chance to tell others about your discoveries. While some scientists will have seen your publications, it is likely that many will not have read your work carefully—there are just too many publications for everyone to keep up on everything, and we tend to focus on just the most relevant work from other groups. Also, the key points of a finding tend to get diluted in a paper, and a lecture provides an opportunity for you to distill down your recent publications into 3-5 bullet points that are important for everyone in the field to know. In addition to telling others about your work, you can also get feedback from leaders in the field during the question period after your seminar or during your poster presentation.

Attending Meetings: Practical Advice

- Wear comfy shoes, especially at extremely large meetings with thousands of posters and large distances between the dozens of seminar rooms and exhibit halls.

- Ask questions, preferably good ones.
    Appropriate times to ask questions and discuss science
        - On the bus to the conference from the airport
        - After a seminar
        - At breakfast, lunch, or dinner
        - During free time
        - On the shuttle bus back to the airport

    Inappropriate times to discuss science
        - In the restroom
        - On the dance floor
        - In the hot tub
        - When you are drunk

- Network—a politically correct term for "party."

Technically, networking means making contacts, not just partying, but lately the "cocktail hour" at conferences has been replaced with "networking hour." A casual exchange of names and pleasantries is not enough—although it is a good start, you need to get to know people on a deeper level. Ask them about their research. Tell them about yours, and bring up any ideas for potential collaborative projects.

Giving a Seminar

When you were a student and post-doc, it was a great learning experience to present a poster at a meeting—the informal discussion with others in the field was helpful. It can still be useful to present a poster at a meeting once you are heading your own research group, but it is also very important to give talks at meetings, thereby reaching a larger audience than possible with a poster presentation. But how do you get invited to give a seminar at a meeting? You can either wait for the phone to ring (or the e-mail invitation to arrive), hoping that your published discoveries and network of friends will be enough, or you can take an active role. The annual meetings of many scientific societies have a mechanism by which members of the society can propose ideas for symposia. This can be an excellent way for junior people to become better known—by being both a speaker and a symposium organizer. Societies often encourage assistant and associate professors to take an active role in the organization of sessions. Another approach to give a talk is to write the organizers of a meeting and request time to present your data. However, this should only be done if you really have a new and important story that would be appropriate for the meeting. Many meetings select some of the talks from the abstracts submitted for the poster sessions, and if you have submitted something that is hot (and write your abstract well) you will likely be selected, without additional effort on your part.

Tips for Giving a Great Seminar

Many years ago it was possible to be a highly successful scientist without being able to give a decent talk. For example, Alexander Fleming was considered a poor speaker, which may partly explain why it took many years for Fleming to interest others in helping him study penicillin. But in modern times, it is very important for your career to be able to give an exciting seminar.

It is often said that practice makes perfect. While practice is certainly useful when giving a talk, and junior scientists should always practice beforehand, too much practice can make a talk boring to the person giving it (unless also a good actor). And, if the person giving the talk is bored, the audience will likely also find it boring. So, definitely practice in front of an audience who can give you constructive feedback but don't get so tired of giving your talk that you lose the excitement—it is more important to be exciting than to have a slick presentation where every word is carefully planned. Never read from a script—this is virtually guaranteed to put your audience to sleep.

Some key points in giving an effective lecture

- Know your audience and target your level appropriately. If a mixed level of senior people and junior people will be present (or senior people in other fields who may not be familiar with your area), try to present something for everyone—go into details, but also have some slides that simplify the system and conclusions.

- Emphasize the basic questions that your experiments were designed to address and why these questions are important. Scientists love a good mystery (as do most people). Have a slide in the early part of your talk that simplifies the system and clearly states the questions you are asking.

- Include a summary slide at the end of your talk that answers the questions you raised in your introduction (or at least most of them). Emphasize the main points you would like the audience to remember.

- Keep to the time limit. Nobody enjoys it when a speaker goes far over the limit; this is not fair to the other speakers in the session who may have to reduce the length of their own talks in order to keep the session on time. Also, if you speak too long there will be less time for questions, and you'll miss an important opportunity to get feedback on your project.

- Give appropriate credit to others. If you think you may forget to show your acknowledgment slide at the end, include the names of collaborators at the beginning or during your seminar.

- Be excited and exciting. Over time, if you make the effort to improve your lecture style, you can learn to give a great talk and avoid putting the entire room to sleep.

While the Whole Room Gently Sleeps
(to the tune "While My Guitar Gently Weeps" by George Harrison)

I look at you all, when I am up there speaking,
While the whole room gently sleeps.
I try to make noise, even shouting and shrieking,
Still the whole room gently sleeps.

I don't know why you find it boring,
And you're all snoring and snide.
It's my best work that I'm presenting,
And now you're denting my pride.

I look at you all fast asleep no one stirring,
While the whole room gently sleeps.
I show new results from which you should be learning,
But still the room gently sleeps.

I don't know how I can excite you,
It's not polite to sleep.
I don't know why you snooze so soundly.
My works profoundly deep!

I look at you all fast asleep no one moving,
While the whole room gently sleeps.
I've answered life's mysteries, as I am proving!
Still the whole room gently sleeps.

Organizing Scientific Symposia

As you move up in your career you will find opportunities to organize sessions within larger meetings, help out on program committees, and finally organize an entire meeting (with the help of a program committee). Although these activities take quite a bit of time and effort, they can be rewarding experiences.

There are ultimately three types of successful meetings. One type is the large society meetings that cover all aspects of a field. These are always organized by committees and usually broken up into subfields or themes. The other two types are small meetings that are either horizontal or vertical sections through a field. Horizontal meetings cover a certain topic in a variety of systems. For example, a meeting on peptides includes everything from chemical synthesis of peptides and peptide analogs to naturally occurring peptides in plants, animals, and bacteria. The link between these diverse areas of biology is the type of molecule being studied. In contrast to horizontal meetings, vertical meetings examine a system at many different levels. For example, meetings focused on a particular disease will often include a range of scientists using different techniques to explore the disease process, from clinical studies to molecular analysis. All three types of meetings provide opportunities for interactions among scientists who share a common bond but also have different approaches—these interactions can be very productive.

To organize a meeting, you need to create a balanced program, get funding to help cover the expenses, and spread the word about your meeting so that you have an audience. On one hand, you want to invite famous scientists who will draw an audience. On the other hand, famous scientists are usually also very busy and may not stay for the meeting. And, if they don't draw an audience, there is the danger that your meeting will have a small audience if the stars all leave after their seminar. It is optimal for most meetings to have a diverse group of speakers, including well-known stars (who may not stay) as well as a core group of senior and junior people in the field who are likely to stick around for the entire meeting, and who may also bring one or two lab members with them.

One of the hardest parts of organizing a meeting is figuring out what will be hot topics many months or even a year in the future. It's easy to scan the literature and see who has published exciting research, but this doesn't mean that these scientists will have another exciting story at the time of the meeting. Most meetings are scheduled 6-12 months in advance, and you want the speakers to include a fair amount of new data (either recently published, or about to be published). If you know people well, you can write them to ask what they're working on and what they think will be hot a year later, at the time of the meeting. I've done this and gotten some quite unexpected responses—people frequently move into new directions and there's no way to know without asking them.

The larger your meeting, the more help you will need with organization. A diverse program committee provides a broad perspective. A large fundraising committee is also important. In my experience, the success rate of fundraising requests depends on personal contacts, and so, a larger committee usually means more contacts.

In addition to the above considerations, you may have to come up with an introduction for each of the speakers. For this latter topic, consider the following template.

Mad Libs® for Scientists: Introduction for a Speaker

Remember Mad Libs®, that whacky party game you played back in grade school? Liven up your symposium with the following Mad Lib® for scientists. It's very simple. First, ask the speaker or audience for words (nouns, adverbs, dates, times, etc) to fill in the following text without explaining what the words are for. Then, read the following text, inserting the appropriate word.

It is a great pleasure to introduce __name__. I have known name ever since we first met on _date_ at precisely __time__, a moment firmly etched in my memory because of the __feeling__ it has caused. Since then, we have repeatedly bumped into each other at meetings. You see, _name_ is rather clumsy and is always bumping into things like __noun #1__, __noun #2__, and even __noun #3__. __Name__ has had a stellar career. I mean that in a literal astrophysical way. When __name__ was young, he/she was mostly hot gas, but now after growing older, he/she has become a burnt-out dense-core __noun#4__ drifting __adverb #1__ through the universe. His/her ideas are revolutionary. Also literally. They are truly revolting, as I'm sure you will agree once you have heard them.

  __Name's__ research has largely focused on __molecule__ in the __body part__ of the __organism #1__, which __name__ claims is a model for __human disease #1__. Despite working in a completely unrelated field, __name__ has led to a breakthrough in the treatment of insomnia. I'm sure you will agree after you hear the talk. Hopefully someone will wake us all up at the end of the __adjective__ seminar.

  In summary, there is nothing that __name__ can't do. Then again, there's nothing that he/she can do either, so let's call it a tie. Without further ado, I present __name__ who will now give a seminar on his/her latest work .

Keeping the Seminars on Time

  Scientists generally love to talk about their work. This is understandable. We work long hours and are excited about our results, so it is natural to want to tell others about our discoveries. It is hard to sum up several years of work by a group of students, post-docs, and other trainees in a 20 minute talk. It is common for the speakers to go over their time limit, and most meetings include a little extra time for discussion at the end of talks as well as during breaks between sessions. But even then, every meeting seems to have a few speakers that seem oblivious to the time limit and go so far over time that the session chair has to intervene. Subtle hints like a timer, bell, buzzer, or red light work for some speakers but others are still impervious and keep on talking. More overt hints are needed, and while tasers, 2000 volt cattle prods, and tear gas would likely be effective, less dangerous methods are preferable. What we need are universal symbols, much like those used in American football. The following are suggestions for signals to be used by session chairs when speakers exceed their time limit.

Chapter 12
The Thrill of Teaching

One of the fun parts of a career in academia is the chance to pass your knowledge on to the next generation of eager young minds, although not everyone perceives this as fun. Especially if the "eager young minds" only care about what's going to be on the exam and aren't interested in really learning the subject. It is also not fun if you have to teach things you don't find exciting. Teaching will take time away from your research, especially if you don't know the subject and have to learn it.

At some colleges and universities, teaching can be the most important determinant of whether you will be promoted—even more important than research! At other places, teaching is just something you have to do, not necessarily well. However, there's no point in doing something badly—you won't enjoy it, and neither will your students. You should try to do a great job at it, regardless of whether teaching is a high or low priority at your institution. Some of the things it takes to be a great teacher do not require much extra time. On the other hand, it does take quite a lot of practice to be a great teacher. Therefore, don't give up if most of your students walk out during your first lecture (as happened to me). Instead, keep trying to improve—you may even win a teaching award (also as happened to me 22 years later—it was quite a transition from my first pathetic attempt at a lecture!).

I find it odd that to teach kindergarten, people need to take classes on education but to teach college, graduate school, or medical school there is usually no formal training in education. While some training programs for graduate students and post-docs offer courses on education, this is not common. Instead, most of us learn on the job—trying out different teaching styles until we find what works.

What is Teaching?

When asked to teach for the first time, many scientists think that it's the same thing as giving a lecture to colleagues at a scientific meeting. While there are some general similarities, there are many differences between the two. Teaching is not merely standing in front of a classroom and lecturing. If nobody is learning, you're not teaching! And how does one define "learning"? There is short-term learning, as measured by exams given within the course. In some cases, there are standardized tests that require longer-term learning (i.e. board exams for medical students, qualifying exams for graduate students). And then there is very long-term learning—remembering for life. This is the hardest to measure, but the most important goal for a teacher.

Another related issue regarding the definition of learning has to do with the definition of memory. Does learning mean spitting back facts on an exam, or truly understanding and being able to use those facts to solve problems? Or is learning a combination of these two: remembering facts and understanding principles? While exams often focus on memorization of facts, this will be useless later in life unless one has a real understanding of the principles. For example, the mathematical term "pi" is both a fact and a concept. We must remember that pi is approximately 3.14—that is a fact. But we also have to remember the concept—pi is the circumference of a circle divided by its diameter, and is useful in calculating the area of a circle, the volume of a sphere, and other things. We need both the fact and the understanding. One without the other would be useless. Understanding the concept without knowing the value would be as pointless as knowing the value without understanding what it is used for. On the other hand, one can easily look up the value of pi (and with more precision than most people can remember it), and so it is more important to understand concepts than facts, especially facts that are easily found.

Teaching Styles

Before discussing specific things that effective teachers tend to do, it is important to stress that there are many different styles of teaching. In my role as the director of a medical school course, I have seen a range of successful styles. One highly-rated lecturer shows a large number of powerpoint slides that complement material presented in the course syllabus. Another highly-rated lecturer shows no slides at all, and just discusses figures and tables in the syllabus, occasionally drawing on the chalkboard. Another highly-rated lecturer just walks around the room and asks questions, waits for an answer, and then discusses the answer. Each style works for the personality of the lecturer and the subject matter.

Although it is important to find your own personal style, you also have to consider the subject you have to teach and the level of your students. Are you expected to give a large amount of information that students need for a standardized test? If so, you have to focus on facts and shouldn't spend too much time on stories. On the other hand, if your objective is to get students to think, then showing a hundred slides is probably not the best plan (not that 100 slides is ever a good idea).

There is also a difference between adult learners and younger students. Much of what is written on education has to do with young students, which require a strong focus on classroom management.[20] Older students such as those in college, medical school, and graduate school cannot be told to do something that they don't want to do—if they want to walk out of the class, they can (and will). Adult learners tend to be self-guided, problem-centered, and results-oriented. Therefore, the style of teaching needs to be appropriately matched to the age and level of your students.

---

[20] Teach Like a Champion: 49 Techniques that Put Students on the Path to College by Doug Lemov (2010).

Be an Effective Teacher

What does it take to be an effective teacher? In the bestselling book <u>Blink</u>, Malcolm Gladwell described an amazing study in which short video recordings of teachers were used to evaluate teacher effectiveness. According to Gladwell, a person watching six seconds of video, without sound, "will reach conclusions about how good that teacher is that are very similar to those of a student who has sat in the teacher's class for an entire semester."[21] If true, this would be remarkable—it would mean that being an effective teacher is communicated entirely non-verbally and is so obvious that just six seconds of viewing is enough to spot. And, this would imply that to become an effective teacher, one needs to focus on these non-verbal cues, whatever they are.

But is this what the study actually showed? The paper cited by Malcolm Gladwell was a study by Nalini Ambady and Robert Rosenthal published in 1993.[22] The people viewing the video were not asked to rate overall effectiveness, but instead were asked to rate the teacher on a number of individual points: activeness, enthusiasm, competence, honesty, professionalism, and several more. Some traits, such as enthusiasm, could be accurately discerned from the short viewing and showed a reasonable correlation with student ratings of the teacher's overall popularity. However, most traits that were evaluated by the viewers of the short videos did not show a statistically significant correlation with the student ratings. Finally, the study did not actually measure whether the students had learned anything from the class, but only measured how

[21] <u>Blink: The Power of Thinking Without Thinking</u>, by Malcolm Gladwell (2007), pg 13.
[22] Ambady, N., & Rosenthal, R. (1993). Half a minute: Predicting teacher evaluations from thin slices of nonverbal behavior and physical attractiveness. Journal of Personality and Social Psychology, 64, 431-441.

popular the teachers were based on student ratings. In other words, the study only showed that being regarded as a popular teacher by students correlates with some behaviors that can be discerned by a group of people watching a silent six second video clip. Popularity is not the same as actually being an effective teacher.

I have sat through many lectures by experts, seen many student evaluations of these experts, and also seen how well students did on exam questions covering various parts of the course. If students performed badly on the exam questions for a particular section, they always gave poor evaluations to the teacher. But in some cases, students also gave poor evaluations to teachers even when they did very well on those parts of the exam. In other words, the students learned the material but didn't think the teachers were great. Was this an effective teacher? The students clearly learned, but weren't happy with the teacher.

Ultimately, you want to teach the students <u>and</u> be considered a great teacher—the student evaluations are often seen by the promotion committees. Based on my experience as a teacher and course director, the following are suggestions to be a great teacher who educates the students and is recognized as an effective teacher by the students.

1. Have clear learning objectives

Before your lecture, you need to decide the key points that you want all students to learn. These objectives should be clearly communicated to the students during your lecture. The objectives of each lecture should not be so broad that they are meaningless (i.e. to learn biochemistry), but instead should be reasonably specific (i.e. to learn the different classes of proteases and understand their mechanisms).

It is a good idea to create exam questions when preparing your lecture as this will help you focus on the key points you expect the students to get out of your lecture.

Quite often, students view lectures as preparation for the exams, and often irritate professors by asking "will this be on the exam?" You don't have to tell them what will be on the exam, but by thinking about exam questions when you prepare the lecture, you will have the same perspective as your students. As a side point, writing exam question is not trivial and even experienced teachers find it helpful to consult with others. Because students will judge your teaching ability by whether you prepared them for the exam, coming up with fair exam questions is an essential part of the teaching process.

2. Have high but realistic expectations

Push your students to learn as much as possible, while keeping in mind the goals of your students. What *do* they really need to know? For example, in a medical school course on pharmacology, do the students need to memorize every drug used, or just a representative group? Or none at all, because they can always look it up later in life when they need to prescribe it? Before they can be doctors, students have to pass standardized board exams, and for this they need to know some details. In considering expectations, be aware that your class is likely to range in background and ability. You need to teach on multiple levels so everyone gets something out of it.

3. Inspire your students to want to learn the subject

Enthusiasm is one of the traits that correlated with perceived teacher effectiveness in the study mentioned in the introduction to this section. To be an effective teacher, you have to motivate your students to become interested in the subject. Some students will probably find the topic interesting without much effort (especially if you are teaching an elective course that they choose to take), but your goal should be to get the entire class excited about the topic.

If you have been assigned to teach something you found extremely boring when you where in school, don't inflict the same punishment on your students by teaching the way you were taught. Try to find something exciting about the topic—most things have some interesting aspects—and include these ideas in your lecture.

4. Make learning fun

Although your primary job is to explain the subject, and not to perform stand-up comedy, it helps to include things that make it fun. Carefully planned visuals—slides that are both scientifically accurate and beautiful—can spice up a lecture. Consider telling stories to illustrate the points you are trying to make, providing that the stories are short and relevant.

A sense of humor can be very effective if used properly. Never make fun of students. Be selective in using humor to avoid distracting from the main points. The effective use of humor in teaching is the subject of an entire book, which I highly recommend.[23]

5. Involve your students

Active learning is much more effective than passive learning. This is succinctly summarized by Confucius in a quote commonly translated as "I hear and I forget, I see and I remember, I do and I understand." Keep in mind that teaching is not just imparting information—you also want to give your students a new way of thinking about things that

---

[23] Professors Are from Mars, Students Are from Snickers: How to Write and Deliver Humor in the Classroom and in Professional Presentations by Ronald A. Berk (2003)

can hopefully be used later in life. For this, active learning is ideal.

Hands-on workshops and labs involve active learning, while most lectures are passive learning. But it is possible to bring an element of active learning to the lectures by asking questions throughout the lecture. This is easier to do with small classes, but I've done this even with a class of 180 students. The questions can be rhetorical ones where you only pause briefly and then explain the answers, or they can be real questions that you ask. If the latter, be sure to wait until a student answers (hopefully voluntarily; students usually don't like to be called on and forced to answer a question). If the answer is wrong, be encouraging and give a hint. If nobody volunteers to answer, the question is either too hard (and you need to give a hint) or it is so easy that everyone knows the answer. Unlike young kids who eagerly raise their hand if you ask something that everyone knows, older students usually hesitate to answer obvious questions.

6. Be consistent on what matters, flexible on what doesn't

Consistency includes fairness to the students as well as to the subject. I have heard some lecturers go off at length on a tangent that is their own belief, not one accepted by most people in the field. It is not fair to pass this off as knowledge, and if mentioned at all, it should be accompanied by a disclaimer that the ideas are not widely accepted.

Flexibility is needed to take into account questions that arise, or if it becomes clear that many students don't know basic things that you assumed were taught in previous courses.

7. Learn a little more about the subject than you need

If you don't already know the subject you are expected to teach, then obviously you have to learn it. But how much do you have to learn? It's a good idea to know more than just enough for your lecture—you want to be able to answer some questions. But you don't have to be an expert on the subject—it is OK to say "I don't know" for an occasional question.

Experts do not always make the best teachers. In my experience, some of the worst lecturers were experts who knew the subject forwards and backwards but just couldn't explain it properly at the appropriate level for the students. It doesn't hurt to be an expert, as long as you focus on realistic learning objects (see points 1 and 2) and don't try to tell the students every single detail. It is most important for you to fully understand the concepts of the material and be able to explain them, and not as important for you to remember all of the details.

Teaching versus Research

Most of the ideas listed above do not require much extra time to implement. But point #7, learning a subject, can take considerable amounts of time. If you are at a school where you are expected to do research and obtain funding, try to avoid spending more time than necessary in the beginning stages of your career as a teacher.

The New Assistant Professor Who Could

*O*nce upon a time in a magic land there was a medical school that taught good little girls and boys how to be doctors. This teaching included courses in biochemistry, physiology, and pharmacology, among many others. Now one day the professor who taught the pharmacology course got an incredible job offer from a big university and he left quite suddenly. Everybody at the medical school was stunned. They asked, "Who is going to teach pharmacology to the medical students?" The chairman of the department said, "I'm too famous to teach the little medical students. Go ask Professor Deadwood—he has tenure and lots of spare time." But Professor Deadwood said, "I'm too old and tired, and besides I don't remember any pharmacology, or much of anything!"

The medical students were sad. "Who's going to teach us pharmacology?" they cried. Just then, a new assistant professor joined the faculty. He had been recruited to the department because of his outstanding research record and although he had no teaching experience, none of the recent recruits ever did. When he saw the sad little medical students he said, "I'll teach the course," not realizing how difficult this really would be.

The new assistant professor immediately began to read about all the subjects of pharmacology: anticholinergics, anti-arrythmics, anesthetics, antihistamines, anti-inflammatory agents, antibiotics, antipsychotics, and much more. It was an uphill battle, trying to learn everything, prepare lecture outlines, and fend off the chairman who constantly inquired as to how his research was coming. As it got closer to the start of his 48 lectures, the new assistant professor was feeling even more hopeless. But, to avoid the bleak reality he kept repeating over and over, "I think I can, I think I can, I think I can." It became his mantra which he would recite while driving to work, while in the shower, and even while sleeping, "I think I can, I think I can, I think I can."

Once the lectures started, the new assistant professor began to build up steam. With each lecture completed, he was one step closer to the end. Pretty soon, he had given all 48 lectures in addition to a few extra case conferences and several small group discussions. Even though the department chairman was very upset that he hadn't gotten his research program up and running, the new assistant professor was satisfied that he had made all the good little medical students happy. That is, until the course evaluation questionnaires were collected. "Terrible!" said some of the students. "No good!" said the others. "He didn't tell us what was going to be on the test," they all whined in unison. They were not happy little medical students at all, but deeply upset and resentful medical students.

Epilog:

The new assistant professor gave up teaching, focused on research, and became a world famous scientist. The good little medical students all passed the medical boards because they had actually learned a great deal of pharmacology in their course. They grew up to be rich little doctors, at least until the Health Maintenance Organizations forced down doctor's salaries in order to generate a hefty profit for the investors.

Moral (pick one):

Where there's a will, there's a way out.

Don't bite off more that you can eschew.

A switch in time saves your job.

Succeed in Science

Chapter 13.
Science and Family

Most of the serious books on career advice do not mention the issues of family life, especially the subject of kids. Maybe this is considered to be incompatible with a professional career and that to raise a family one needs to have a non-scientist spouse take on the majority of the child rearing. But it is possible to be an active parent and a full-time scientist, and there are many cases of couples where both parents are successful scientists.

It is said that there are many rewards of being a parent. For example, you qualify for a larger tax deduction. But more importantly, there is the reward of having accomplished a basic and fundamental aspect common to all life forms. Many scientists don't think they can afford the time away from the lab. But it is possible to spend time with kids and do something related to science such as writing, reading, and reviewing papers and grant applications, and just thinking about things. In my experience the time I spent while interacting with my kids was a great break from the intensity of science. Sometimes the best solution to a problem appears when not actively thinking about it. Also, it is possible to combine science with parenting duties. When kids are very young, they just need an adult around for an occasional feeding, diaper change, and hug—the rest of the time they sleep, poop, and/or stare at nearby molecules of air. While it is good for them to hear the sound of your voice, it doesn't really matter to them if you are reading aloud the old children's rhymes or practicing a seminar you have to give. You can also amuse them by reading excerpts of the grant application that you are working on. If they smile, it is good. If they fall asleep, consider revising to make more exciting. And if they cry or throw-up, there are major problems and you must rewrite!

Finding a Mate

Unless you have perfected human cloning, the first step in raising a family is to find a mate. This can be hard if you spend all your time in lab. Consider placing an advertisement in an on-line dating agency. Be careful what you write—you don't want to attract the wrong person. Consider how proteins find other proteins to interact with.

Postings in the Protein Dating Agency

Single protease, into heavy metal (zinc), seeks compatible protein that can fit my cleft. I'm not into long-term relationships, but one night with me and you'll never be the same again. If you're looking for cleavage, you won't be disappointed.

Bifunctional protein, recently detached, wants to meet proteins who know how to get around. I am attracted to either hetero- or homo-dimers, so let me undo your leucine zipper and squeeze into your active site.

Shapely lectin seeks mature glycoprotein for a stable relationship. Be my sweet talking sugar daddy and I'll stick by your side. Are you mannose enough for me?

Nitric oxide synthase seeks inducible guanylate cyclase who just can't say NO. I got the Viagra if you got the time.

Recently separated subunit looking for transmembrane protein who knows how to caress my amphipathic helix. No peripheral membrane proteins with an attitude.

Activated heptahelical receptor protein wants to turn you on. If you're the right match for me, I can stimulate your G-protein and split your alpha and beta-gamma subunits wide apart. Meet me at the coated pit.

ER resident looking for young unfolded proteins who are not afraid of trying new positions. I got plenty of time, experience, and energy (ATP). I'll be your chaperone if you'll let me get under your beta-sheets.

Single subunit protein seeks a DNA-binding protein. If you can stimulate my helix-loop-helix with your zinc finger, let's find the groove and turn on transcription together, baby. Meet me at the E-box. No transcription repressors, please.

Topoisomerase looking for DNA to unwind with. If your idea of fun is doing the twist, let's meet and see what unravels. There's nothing too kinky for me!

Aging extracellular matrix protein looking for a stable relationship. Let's spend our golden years together with candlelight dinners in the caveolae, romantic walks on the basal lamina, and a twilight cruise to an exotic lysosome.

The radiocarbon dating technique.

Falling in Love, or Something!

You finally meet someone special and get a strange feeling inside—something you never experienced before. But, being a scientist, you need to be sure that this is love, because after all, it could be a number of other things.

Is This Love?

Is this love or is it
Pregna-4,20-diene-3,6-dione
Or another pheromone that you secrete
To stimulate my vomeronasal system
Which gives the illusion of love
Or some kind of attraction
But it's really just chemistry.

Is this love or is it
A calcium current in the neurons within my nucleus
    accumbens
That leads to a massive release of dopamine and
    other neurotransmitters
To activate my entire basal ganglia
Which gives the illusion of love
Or some kind of attraction
But it's really just neuroscience.

Is this love or is it
A conditioned positive reinforcement reflex
Of a primordial psychosexual procreation impulse
From within my subconscious collection of repressed
    memories
Which gives the illusion of love
Or some kind of attraction
But it's really just psychology.

Is this love or is it
A predetermined genetic program by which my DNA
    can replicate
With you merely the vehicle that my DNA has
    selected
To spread over the planet like a virus
Which gives the illusion of love
Or some kind of attraction
But it's really just genetics.

Is this love or is it
"A" if all of the above
Or "B" if 1, 2, and 3 are correct
Or "C" if 2 and 4 are correct
Or "D" if only 4 is correct
Or "E" if none of the above
And it really is love
Or maybe just a stomach virus.

"Yes, love is a potent drug indeed, Miss Cruickshank....
But I still don't think we can analyze it using
Gas Chromatography..."

Marriage—the Pros and Cons

In many professions, it is easy to live together without getting married. While this is an option for scientists too, most science jobs come with great benefits (health care and other perks) and these will usually only cover your partner if you are married. On the other hand, the time and effort to get married means less time for your science research.

If your partner is also a scientist, consider combining your honeymoon with a scientific meeting (this only works if your spouse is in the same field and wants to go to the same meeting). My wife and I actually did this—in our case we met at a scientific meeting, began dating (long distance at first), eventually were post-docs together in the same lab, and then after getting married, went on a honeymoon to an exotic place where we attended the annual conference of the same society of which we first met. It was an efficient use of time—rather than lie on the beach and doze in the hot sun, we sat through seminars where we could doze in darkened auditoriums and avoid the sunburn. However, not all couples will find this romantic. If your fiancé wants a large wedding and long honeymoon, you just have to take time off from science and go along with the plan. Hopefully you can find time to check your e-mail during your honeymoon, but it is not advised to do this during the actual wedding ceremony or anytime on your wedding night.

Logic doesn't control who we fall in love with, but logic can play a part in deciding whether to get married. If you are not sure if you should marry another scientist or a "normal" person, take this simple quiz.

Quiz to Test if You Should Marry Another Scientist or a Non-Scientist

During my free time, I like to

  a) relax and read ordinary books and then discuss them with other people.
  b) read scientific articles and then discuss them with other people.
  c) what free time?

Most of my friends are

  a) non-scientists, and we talk about a broad range of topics, from movies to sports.
  b) other scientists, and we talk mostly about science.
  c) laboratory organisms.

My idea of a dream vacation is

  a) a week at a beach resort with days spent swimming, sailing, and reading novels, and nights spent dining in exotic restaurants and dancing to lively local bands until the early morning hours.
  b) a week at a scientific meeting with days spent listening to seminars and nights spent talking science with colleagues until the early morning hours.
  c) a week in lab with no distractions.

Scoring: If you chose "a" for all answers you would be happiest with a non-scientist. If you chose "b" for all answers you're destined to marry another scientist. If you chose "c" for all answers, get back to lab and stop wasting your time reading this book or even thinking about getting married.

Pros and cons of marrying another scientist in your field

Pro: You can help each other with designing and interpreting experiments, writing papers and grant applications, and sharing reagents.

Con: In addition to normal things to argue over, you can also argue about the best experiments to do, whose hypothesis was proven wrong by the data, and who should be first author and senior author on your collaborative manuscripts.

Pro: You can attend the same scientific conferences, thereby saving on room costs and allowing you to spend time together.

Con: If you have kids, who is going to take time off from the conference to spend with them? Also, if you spend most of your time at the conference with your spouse, you will spend less time networking with other people, which is one of the reasons to go to meetings in the first place.

Pro: Two heads are better than one! You can be more efficient and productive by working as a team.

Con: Who's going to hire two people for one job? Most departments want a balanced group of scientists pursuing different problems, and two scientists in the same field may not be able to get jobs at the same university.

Pros and cons of marrying a scientist in another field

Pro: You can learn about your spouse's field, thus expanding your knowledge base.

Con: This probably won't help much in your own research.

Pro: Unlike marrying a scientist in your own field, a scientist in another field will never argue with you about experimental details.

Con: Instead, you can argue about whose field is more scientifically advanced and/or socially relevant.

Pro: If you go along to one of your spouse's scientific conferences, it will be a real vacation for you.

Con: A vacation alone, that is! Your spouse will be so busy with the meeting that he/she won't have any time to spend with you.

Pros and cons of marrying a non-scientist

Pro: When you arrive home late at night, dinner will be on the table.

Con: Although dinner will be on the table, it will be cold and you'll be eating alone because your spouse went to bed hours ago.

Pro: It is unlikely that you and your spouse will ever argue about details of scientific experiments.

Con: What on earth will you talk about once you exhaust all the common subjects after several years of marriage?

Pro: You can lead a normal life, or at least pretend to be normal.

Con: Deep down inside, you can never shake the feeling that your spouse considers you a nut for obsessing over something that nobody else can comprehend.

Until Bob entered her life, Lisa never even knew the meaning of the phrase 'acute multiple drug-resistant infection by transgenic Staphylococcus aureus.'

Funny now to think back... Somehow all these little things change when you date a careless microbiologist...

Having Kids and Keeping Your Job

So you fall in love and get married. It's not really so different from single life, at least not yet. But the next step—having a kid (or kids)—is a huge change in your life-style. The mere thought may seem incompatible with a career in science, which typically requires 50+ hours of work a week. On the other hand, a week has 168 hours, so if you manage your time wisely (and don't sleep much) it is possible to be a successful scientist and an active parent.

In my case, when my children were very young I stayed home from work one day each week and my wife did the same (and the other three mid-week days we found relatives / friends / neighbors / baby sitters to watch them). The time I spent at home with my young children was quite productive—they slept most of the time and I was able to work on manuscripts and do other deskwork. At times when I absolutely had to do experiments on my assigned parent day, I would bring the young child into lab, set him or her up in a stroller in my office, and run back and forth between the lab and the office to check every few minutes. Ultimately, it all worked out and I was able to juggle the time spent with the children and time spent on science. The only time it didn't quite work out was one particular late night. My daughter went through all of the diapers I had brought along, and in desperation I wrapped her up in the adsorbent pads used to cover lab benches (I did use new ones that had not been on a lab bench!). All was fine until the end of the day when we were leaving. A colleague saw us in the building lobby and went to pick her up out of the stroller. The make-shift diaper fell apart and it was quite embarrassing for all parties involved. After that, I packed twice as many diapers and even kept a few extras tucked in my desk drawer. They may still even be there, so if you need one and are in the area, stop by!

Children's Rhymes with a Scientific Twist

Let's face it—most stories and poems for children are pretty stupid! We hear these things so many times that they become firmly etched in our memory. But except for the "Alphabet Song," which is actually pretty useful for learning the alphabet, most children's songs and rhymes serve no purpose. If we are going to commit something to memory, shouldn't it be something useful? Why not change children's rhymes to emphasize important scientific points? Perhaps this would cause the next generation to have a greater appreciation for science. Here are some examples.

Little Boy Blue

Little boy blue come blow your horn,
The sheep's in the headlines for being born.
And where is the scientist who cloned the sheep?
He's preparing his Nobel Prize acceptance speech.

Hickory Dickory Dock

Hickory dickory dock,
The mouse ran up the clock.
The clock struck one, and down he run,
And the neuroscientists observing this activity wrote a 15 page paper describing it as a novel aversive condition avoidance behavioral test.

Little Ms. Muffet

Little Ms. Muffet sat on her tuffet
Experimenting on DNA.
She created a spider, which made golden fiber,
And she patented the process right away.

Top Ten Complaints from Children When Both Parents are Scientists

10) You can't understand a single word said at the dinner table once your parents start talking about what they did at work that day.

9) No matter what you ask, the answer is always "well, that's a really great question" followed by a ten minute lecture.

8) When you think about all the times your parents have talked about genetic engineering, you can't help but wonder if you are playing with a full set of genes.

7) You have to get an A+ in science and math or else your parents will be disappointed and ask you what went wrong.

6) Having to spend a week with Grandma hearing about her hernia operation while your parents are off at conferences.

5) Interesting dinners when your parents decide to experiment in the kitchen.

4) Having to listen to conversations with the words "thereby," "wherefore," and "thusly."

3) The two months before a grant application deadline.... and the eight months afterwards.

2) Your friends get to watch cool TV shows but all you can watch are science and nature documentaries such as "the feeding habits of the mongoose."

1) You'll always have the feeling, even after years of therapy, that being "excellent" is not quite good enough.

Summer Science Camps

Every spring, millions of parents are faced with the problem of what to do with their school-aged kids during the long summer break (which didn't seem long enough when we were kids, but now seems quite excessive when we are parents and have to find something to occupy our kids for 10+ weeks every summer). Although there are numerous summer camps for normal kids, scientists are somewhat reluctant to send their children to camp where the major intellectual activities involve coordinating late night panty raids or arranging tadpole eating contests. There is a need for camps that meet the special requirements of future scientists. The following is a listing of camps that don't quite exist yet but are likely to appear due to the basic principle that running a camp is a great way to make millions of dollars by working only 8 weeks a year!

Camp O-Wa-Ta-La-Ta-Fun

This camp combines the best of regular summer camp with the fun of science. In addition to standard camp activities such as swimming, boating, horseback riding, team sports, and food fights, this camp has a gigantic science lab where campers learn about physics, chemistry, and the various "ologies" (biology, physiology, pathology, zoology, geology, archeology, and pharmacology). Hands-on demonstrations of fundamental scientific principles are provided. Kids will thrill in the adventure of discovering, all on their own, such important scientific concepts as gravity, fire, and penicillin. The highlight of every camper's experience is our weekly bonfire and sleepout. Gather 'round the campfire for chilling real-life stories such as the basement chemist who inadvertently discovered a neurotoxin and developed Parkinson's Disease, the post-doc who fabricated data and got caught, or the professors who thought they discovered cold fusion and hastily called a press conference rather than attempting to first publish the results in a peer-reviewed journal (or even better, to reproduce them!). This camp is co-ed, of course, for those informal anatomy lessons behind the bushes.

Camp Em-Dee-Wa-Na-Bee

This camp is popular among kids who want to be doctors. Special emphasis is placed on pre-med courses and on science projects (so you'll have something exciting to talk about during your college and medical school interviews). This camp is affiliated with a real live Nobel Laureate who will visit the camp, shake your hand, and then personally write you a recommendation letter for entry into the pre-med program and medical school of your choice.

Camp Wee-Nee-Dah-Ten-Shun

This camp is designed specifically for children of scientists. The major daily activity is support group discussion sessions modeled after those of "children of alcoholics." Discussion topics include, "I know my parents really do love me even though they..." (pick one of the following): (a) packed me off to summer camp for 8 weeks so they could work in lab 18 hours per day; (b) are more interested in reading back issues of scientific journals than my school report on Alexander the Great; and (c) don't always remember my name. Learn key words and phrases to use in conversations with your scientist parents so that they'll include you in the conversation (or rather, so that they'll direct their nightly monologue towards you). Hands on workshops teach special skills needed to live with scientist parents such as how to get attention at the dinner table without resorting to violence, feigning interest in their incredibly minuscule discovery so they'll let you borrow money, and how to tell them that you want to major in art history without sending them into shock.

How we get scientists

Chapter 14
The Grand Finale

In summary, science is an amazing profession. If you are innately curious and want to spend the majority of your time figuring out the unknown, then there is no better job. But to succeed, you need to first find a job opportunity and convince the search committee to hire you, then set up your lab and hire, train, and motivate people to pursue your ideas. You also need to write up your results, get the manuscripts accepted for publication, and obtain grant funding. In addition, you need to attend scientific conferences to keep abreast of new developments and establish yourself in the field. For some positions, you need to learn how to be an effective teacher. All while struggling to raise a family. But despite the long hours, being a scientist sure beats working a real job!

It is often said that it ain't over till the fat lady sings, and so the appropriate conclusion for this book of random advice and humor is an opera that summarizes the life of a scientist.

Il Destino di Scientist: A Tragic Comedy Opera in Two Indecent Acts

Notes on the opera: Originally written in the mid-1800s by Giuseppe Linguini, this opera was lost for many years until being discovered in the seat cushions of an antique Italian sofa. It is uncanny that this obscure Italian composer would so accurately portray the life of a professor in the modern world, especially the details about the grant review process used by National Institutes of Health (NIH) in the last half of the 20th century. It is also uncanny that the musical score is identical to selected songs from various Gilbert and Sullivan operas.

<u>Cast</u> (in order of vocal appearance)
    Lucia, a post-doc ....Soprano
    Alfredo, a professor .... Baritone
    Eduardo, the chair of the department...... Bass
    Stephano, another professor....................Baritone
    Erminio, a grant applicant ....Tenor

Act 1, Alfredo's Office

Act II, A Holiday Inn in Valhalla, home of the Gods and
    Goddesses of NIH

<u>Act I, Alfredo's Office:</u>

The curtain rises. Lucia, a post-doc in Alfredo's lab, is
working on a manuscript. She sings about the life of a post-
doc.

I Am Just a Poor Post-Doc
(to the tune of *I Am Little Buttercup* from <u>H.M.S. Pinafore</u>)

    I am just a poor post-doc, forever-more post-doc,
    Though for real jobs I apply.
    But still I'm a poor post-doc, can't move beyond post-
        doc,
    Post-doc I'll be till I die.

    I've worked hard for ages, despite the low wages.
    No bank account have I at all.
    I am just a P-H-D, that's why my salary
    Is so incredibly small.

    I polish my résumé, hoping to see a day
    When a good job ad I'll find.
    I will apply right away, and with luck they will say,
    "We want exactly your kind."

I'll breeze through the interviews, and then if I choose,
I'd start work the very next day.
They'd pay me six figures, and then in the good years
A bonus would double my pay.

I'll make half a million, and then while I'm still young
The company stock I'd acquire.
And when the shares double, I will have no trouble
At age forty five to retire.

But sadly I must say, the job I have today,
Is still but a poor post-doc, I.
Because there's no company, wanting to hire me,
Post-doc I'll be till I die.

———————————

Alfredo enters with several graduate students and tries to cheer up Lucia by explaining how he got to be a professor. The students take part in the song as the chorus.

The Leader of a Lab-ra-try
(to the tune *When I Was a Lad* from H.M.S. Pinafore)

When I was a lad I went to school
And there I learned about the Golden Rule.
I played on the playground, then a nap I took
And I colored all the circles in my coloring book.
(Chorus) He colored all the circles in his coloring book.
I colored all the circles so carefully that now I am the
leader of a lab-ra-try.
(Chorus) He colored all the circles so carefully that now he is
the leader of a lab-ra-try.

Then to the university I went
Where I learned about the scientific precedent.
I studied all I could about each -ology
And I memorized the phyla in each family tree.
(Chorus) And he memorized the phyla in each family tree.

I memorized the phyla so anally that now I am the leader
of a lab-ra-try.
(Chorus) He memorized the phyla so anally that now he is
the leader of a lab-ra-try.

I studied very hard and they accepted me
Into a first-rate program for a P-H-D.
I found a well-known teacher and wrote down each word
That he uttered in his discourse even if absurd.
(Chorus) That he uttered in his discourse even if absurd.
I worshipped each word with such idolatry that now I am
the leader of a lab-ra-try.
(Chorus) He worshipped each word with such idolatry that
now he is the leader of a lab-ra-try.

After graduation then a post-doc I
In the lab-ratory of a famous guy.
I followed the instructions that he told me
But in spite of this I stumbled on a new theory.
(Chorus) But in spite of this he stumbled on a new theory.
My theory was so brilliant that they soon made me the
leader of my very own lab-or-a-try.
(Chorus) His theory was so brilliant that they soon made he
the leader of his very own lab-or-a-try.

So if you want to lead your own lab-or-ra-try
Then you should ignore scientific history.
Just go and make your very own discovery
And claim that it is true with much opiniatry.
(Chorus) Just claim it must be true with such opiniatry.
Then once it is accepted, so well-known you'll be that
you can be the leader of a lab-ra-try.
(Chorus) Then once it is accepted, so well-known you'll be
that you can be the leader of a lab-ra-try.

Alfredo apologizes to Lucia for not helping her, as he is very busy reviewing grant applications—he was recently asked to serve as a grant reviewer and has spent the last two weeks doing nothing but reading the 23 applications that he was assigned to review. Just then Eduardo, the chair of Alfredo's department, enters along with several other faculty members and administrators. Eduardo sings about the difficult life of a department chair, with all others present in the room participating in the chorus.

A Chairman's Life is Not a Happy One
(to the tune of *A Policeman's Lot is Not a Happy One* from
    The Pirates of Penzance)
(Note: the syllable for the first downbeat of every line is
    underlined)

Chairman: When the <u>fac</u>ulty aren't busy entertaining,
Chorus: entertaining
Chairman: They are <u>ask</u>ing me for higher salaries.
Chorus: salaries
Chairman: They've a <u>tend</u>ency for griping and complaining
Chorus: and complaining
Chairman: That they <u>must</u> have larger lab-or-a-tor-ies.
Chorus: -a-tor-ies
Chairman: When the <u>nev</u>er-ending paperwork does smother
Chorus: work does smother
Chairman: When ad<u>min</u>istrative duties must be done,
Chorus: must be done
Chairman: If it <u>isn</u>'t one thing then it is another
Chorus: is another
Chairman: A <u>chair</u>man's life is not a happy one.
Chorus: Ohhhh! When ad<u>min</u>istrative duties must be
                done—must be done,
                A <u>chair</u>man's life is not a happy one—happy one.

Chairman: When the <u>new</u> assistant professor has problems,
Chorus: -or has problems

Chairman: And he <u>needs</u> help from the university.
Chorus: -versity
Chairman: I must <u>go</u> and beg the dean to give some money,
Chorus: give some money
Chairman: Because <u>that</u> is my responsibility.
Chorus: -bility
Chairman: Certain <u>times</u> I feel as if I am their mother,
Chorus: -am their mother
Chairman: There is <u>al</u>ways something more that must be done.
Chorus: -must be done
Chairman: If it <u>isn't</u> one thing then it is another,
Chorus: -is another
Chairman: A <u>chair</u>man's life is not a happy one.
Chorus: Ohhhh! When ad<u>min</u>istrative duties must be done—must be done,
A <u>chair</u>man's life is not a happy one—happy one.

Chairman: And <u>when</u> a junior member needs a letter,
Chorus: needs a letter
Chairman: I must <u>write</u> to the promotion committee.
Chorus: committee
Chairman: Ex<u>plain</u>ing why this candidate is better,
Chorus: 'date is better
Chairman: Than <u>it</u> appears from his or her CV.
Chorus: her CV
Chairman: And <u>all</u> the time I'm constantly pretending,
Chorus: 'ly pretending
Chairman: That I <u>under</u>stand the work of everyone.
Chorus: everyone
Chairman: The <u>tasks</u> that I must do are never-ending,
Chorus: never-ending
Chairman: A <u>chair</u>man's life is not a happy one.
Chorus: Ohhhh! When ad<u>min</u>istrative duties must be done—must be done
A <u>chair</u>man's life is not a happy one—happy one.

Eduardo asks Alfredo if he is managing the grant reviews. Alfredo responds that it is going very well, and that he has a simple system for reviewing grant applications which he explains in the following song.

I've Got a Triage List
(to the tune of *I've Got a Little List* from <u>The Mikado</u>)

As it seems to be essential that the weak grants must be
found,
To make the triage list, I've got to have a list.
There are some common problems I've encountered
every round.
Not one of them is missed, not one is ever missed.
The grant with figures shrunk so small that there is just
no hope,
That anyone can read them without a good microscope.
The grant that is so sloppy it was written in such haste.
I'm sure that any funding to this group would be a waste.
With errors typographical and many contradicts',
This makes my triage list, I've got it on my list.
(Chorus) He's got it on the list, it's made the triage list.
(Chorus) And I don't think they'll be missed, I'm—sure—
they—won't—be—missed.

There's the grant that complicates things and the others
of this type.
The chronic pessimist, I've got this on my list.
The one that overstates things with a lot of dazzling
hype.
The part-time scientist, this surely won't be missed.
The grant that is so boring, many times it made me snore.
It doesn't have excitement, all the works been done
before.
The grant that's way too ambitious, too much they say
they'll test.
If they do one percent of this, why then I'll be impressed.

The grant with jargon so arcane, I cannot get the gist.
I've got it on my list, yes-yes it's made my list.
(Chorus) He's got them on the list, the nasty triage list.
(Chorus) And I don't think they'll be missed, I'm—sure—
    they—won't—be—missed.

I've got a stack of applications I must winnow down.
To make the triage list, my lengthy triage list.
These are the common problems that will make a grant
    unsound,
And none are ever missed, not one is ever missed.
The grant that's in my area, of which I have much fame,
But in their list of citations, not one includes my name.
And worst of all, the application that has been revised,
The hatred for the reviewers is too thinly disguised.
The writer of the grant is, though it's wrong, so rightly
    pissed.
He surely made my list, no doubt he made my list.
(Chorus) He's got them on the list, the rotten triage list.
(Chorus) And I don't think they'll be missed, I'm—sure—
    they—won't—be—missed.

Act II. A Holiday Inn in Valhalla, home of the Gods and Goddesses of NIH

The scene opens to reveal a large table surrounded by serious looking men and women. Alfredo is among the mortals who have been invited to Valhalla to decide the fate of 137 grant applications. At the side of the room are the Gods and Goddesses of NIH, the program officers of the various agencies, dressed in white tunics. They are feeding from a large tray of grapes and drinking decaf coffee. Stephano, a well-known professor from a big-name university explains to the new reviewers how he reviews applications.

The Very Model of a Reviewer Anonymous
(to the tune of *I Am the Very Model of a Modern Major General* from <u>The Pirates of Penzance</u>)

I am the very model of a re-view-er anonymous
I've information veg-e-table, mineral, and animus
I quote the facts historical in orders categorious
And I can tell at sight when there are flaws that are
    enormious.
I know quite well to recognize experiments preposterous
From trivial to artifacts of scientific importous.
In short in matters veg-e-table, mineral, and animus
I am the very model of a re-view-er anonymous.

I'm very well acquainted too with recognizing pure B-S
When it is clear the applicant has gotten in a sticky mess
Because their data do not validate their own hypothesis,
They will receive from me a score that will cause them a
    lot of stress.
When there are no controls proposed, the error is quite
    big to us,
And otherwise when nothing's clear, and everything's
    ambiguous,
When data in the application is so weak and dubious,
There is no hope that I will give a score that is enthusius.

I'd like to think I've seen it all but now and then I must
    confess,
I see an application that is innovative, more or less,
It's logical and fun to read, the applicant was studious,
But still I criticize it so that I appear the gen-i-us.
Though usually the applications have faults rather
    obvious,
A fatal flaw so very large, the applicant's impervious
And pays no heed to redress all the criticisms previous,
Then triage is the score and even this is not so devious.

I know the diff-er-ence between a laptop and an abacus,

And when the revised introduction is quite truly amicus
Or when it is so false that it is surely platitudinous.
And when the applicant tried very hard to be excluding
us.
I know some phraseology that borders on ridiculous,
And other terms that are so mean they really are
insidious.
"Deconvolution of the data will be very problemous,"
I say in my review although the meaning is quite
vacuous.

I am the very model of a re-view-er anonymous,
Although the roster shows my name and so the job is
perilous.
I do incur the wrath of applicants who are so furious,
They feel my pearls of wisdom are just mean, contrite,
and spurious.
I trash them all from neophyte to professor emeritus.
Although at random times I can appear to be quite
generous.
In short in matters veg-e-table, mineral and animus
I am the very model of a re-view-er anonymous.

The review session begins and Alfredo and Stephano debate
the merits of the first grant application.

See How the Fates Our Scores Allot
(to the tune of *See How the Fates Their Gifts Allot* from <u>The
Mikado</u>)

Alfredo: See how the fates our scores allot,
              If A is funded— B is not.
              Yet B is worthy, I dare say,
              Of more in-no-va-tion than A!
Stephano: Is B more worthy?

Alfredo: I should say,
             It could have more impact than A.

Stephano: Yet A is solid,
             Oh, so solid.
             Aim 1, solid.
             Aim 2, solid.
             Even Aim 3, so solid.
             Always simple, plain as day,
             There is nothing wrong with A.
             Never risky, A-O-K,
             Boring non-creative A.

Alfredo: If I were Council, which I'm not,
             B should enjoy A's happy lot.
             And A should fall in triage hell,
             So that grant B could do well.
Stephano: But A unfunded?
Alfredo: That should be,
             There's no cre-a-tiv-ity.

Stephano: B should be funded.
             So well-funded.
             New ground, really,
             Profound, really,
             Not so sound, yes I agree.
             So condemned to triage thee,
             Wretched innovative B.
             Too far out and too risky,
             Way too cre-a-tive is B.
             Nobel Prize is quite likely,
             Wretched innovative B.

A man in a Holiday Inn uniform who is restocking the urns of coffee becomes noticeably distressed and begins consuming vast quantities of coffee during the preceding aria. When it ends, he rips off his uniform to reveal that he is Erminio, the applicant of the grant that just went down the tubes. Even though Erminio is fatally poisoned by an overdose of caffeine, he is still able to sing a moving aria, pleading with the reviewers to at least give his application a score.

Don't Propose Triage
(to the tune of *Go Away, Madam* from <u>Iolanthe</u>)

Don't propose triage,
Or compose triage.
To impose triage is a blow.
Do not say triage,
Or convey triage.
To display triage hurts me so.

It is rude, triage,
To allude, triage,
Or conclude triage, in your haste.
'Cuz the word, triage
Is absurd, triage
Never heard triage in good taste.

Don't decide triage,
It is snide, triage.
I can't hide triage from my peers.
Do not dare triage.
I can't bear triage.
Be aware triage will bring tears.

I will grieve, triage,
To receive triage.
Please don't leave triage, give a score.

If I get triage,
I'm all wet, triage,
So upset, triage, ever more.

The reviewers are not moved by Erminio's pleas, especially when it becomes clear that he is fatally poisoned by the overdose of caffeine and therefore does not really need the grant anymore. The reviewers and NIH officials sing a moving tribute to Erminio and the life of a scientist.

For I Am a Scientist
(to the tune of *For He Is an Englishman* from <u>H.M.S. Pinafore</u>)

Chorus:	He is a scientist.
Erminio:	I am a scientist.
	Although the funding isn't there,
	It is a problem I will bear,
	For I am a scientist.
Chorus:	For he is a scientist.
Erminio:	For I might have been a dentist,
	Or physician's apprentice,
	Or something quite portentous.
Chorus:	Or something quite portentous.
Erminio:	But despite my strong obsession
	To choose a new profession,
	I remain a scientist.
	I remain a scientist.
Chorus:	But despite the budget crisis,
	And funding paylines barely missed,
	We remain as scientists.
	We remain as scientists.

Erminio:	There is no chance of recovery,
	Without a big discovery!
	I remain a scientist.
	Till death, a sci-i-i-i-i-i-i-i-i-i-entist.
Chorus:	But despite the budget crisis
	And funding paylines barely missed
	We remain as scientists.
	We remain as s sci-i-i-i-i-i-i-i-i-i-entist.

At the conclusion of his aria Erminio collapses, fatally poisoned by the vast quantities of caffeine he consumed. The opera ends with the reviewers placing Erminio's lifeless body, still twitching from the caffeine, in the boxes that hold the discarded grant applications and covering him with glossy photos of his data. As the curtain is being slowly lowered, one of the reviewers comments that it's a good thing the application wasn't given a really bad score or who knows what he would have done.

Acknowledgments

Thanks to my wife, Lakshmi Devi for help with editing (i.e. keeping the really weird stuff out) and for contributing some of the serious advice which inadvertently made its way into this book, which was originally intended to be purely a collection of humor essays. Thanks also to our children who put up with two scientific parents, and are budding young scientists themselves. Thanks are also due to many others who gave serious advice and/or reviewed preliminary drafts of this book (Peter, Julia, Iris, Jakie, Sandy, Wing, Jonathan, Stas, and Jeff). Last but not least, the dozens of sources of inspiration: William (Bud) Abbott, Lou Costello, William Gilbert, Arthur Sullivan, Woody Allen, Dave Barry, the writers of Mad Magazine, and most importantly, all of the people whose manuscripts, grant applications, promotion packages, and/or job applications I have reviewed over the past 25 years.

About the Author and Illustrator/Cartoonist

Lloyd Fricker is a Professor in the Departments of Molecular Pharmacology and Neuroscience at Albert Einstein College of Medicine in New York City. When not leading a research group, writing and reviewing scientific papers and grant applications, teaching medical and graduate students, organizing and attending scientific conferences, and serving on various committees, he spends his time writing humor. His humor pieces (including some in the present book) have appeared in Annals of Improbable Research, Journal of Irreproducible Research, The Einstein Journal of Medicine and Biology, H.M.S. Beagle, and several other journals, as well as youtube. His first book on scientific advice/humor, "How to Write a REALLY Bad Grant Application (and Other Helpful Advice For Scientists)" was published in 2004.

Nick Kim has a background as a Senior Lecturer and local government scientific adviser in Applied Environmental Chemistry and currently lives in Wellington, New Zealand. In his spare time he draws from inspiration as a scientist and creates cartoons. His cartoons have appeared on the web since 1994 (http://www.lab-initio.com/), and in science-related publications such as Annals of Improbable Research, Physics Today, Chemical Innovation, New Scientist, Skeptic (UK), Chemistry in Australia, and New Zealand Science Monthly. Nick is also a contributing editor for the Annals of Improbable Research.